Methods in Physiology

Published on behalf of The American Physiological Society by Springer

Methods in Physiology

This series describes experimental techniques in cellular, molecular and general physiology. Each book is edited by experts in their field and covers theory and history behind the method, critical commentary, major applications with examples, limitations and extensions of each technique, and vital future directions.

More information about this series at http://www.springer.com/series/11781

Jatin George Burniston • Yi-Wen Chen
Editors

Omics Approaches to Understanding Muscle Biology

Editors
Jatin George Burniston
Research Institute for Sport and Exercise Sciences
Liverpool John Moores University
Liverpool, Merseyside, UK

Yi-Wen Chen
Children's National Health System
George Washington University
Washington, DC, USA

Methods in Physiology
ISBN 978-1-4939-9804-3 ISBN 978-1-4939-9802-9 (eBook)
https://doi.org/10.1007/978-1-4939-9802-9

This Springer imprint is published by the registered company Springer Science+Business Media, LLC, part of Springer Nature.
The registered company address is: 233 Spring Street, New York, NY 10013, U.S.A.

Contents

About the Authors

Jatin George Burniston is a Professor of Muscle Proteomics at the Research Institute for Sport and Exercise Sciences, Liverpool John Moores University, UK. He established the first proteomics facility with a specific focus on exercise physiology and published the first works reporting proteomic analysis of striated muscle responses to exercise training. Jatin strives to continue pioneering the field of exercise proteomics and is a proud champion of the application of non-targeted '-omic' research in exercise physiology. He recently combined his expertise in mass spectrometry and proteomics with new metabolic labelling methods using deuterated/'heavy' water to establish the unique 'Dynamic Proteome Profiling' method. Jatin serves on the Editorial Board of the American Physiological Society (APS) journal, *Physiological Genomics*, and is a Fellow of the European College of Sports Sciences (ECSS) and a member of the ECSS Scientific Committee. He also leads the MSc Exercise Physiology programme at Liverpool John Moores and particularly enjoys delivering his module on Molecular Exercise Physiology.

Yi-Wen Chen is an Associate Professor of Genomics and Precision Medicine at the George Washington University and Principal Investigator at the Center for Genetic Medicine Research, Children's National Health System, DC, USA. Her research uses transcriptomic approaches to study molecular pathways in muscles in response to physiological stimuli and pathological conditions. Yi-Wen uses genome-wide RNA profiling to dissect the disease mechanisms of various muscle disorders and to identify molecular mechanisms of adaptive remodelling. In addition, her group uses next-generation sequencing and long-read sequencing to study genetic, epigenetic and transcriptomic changes in muscles using physiological and disease models. Currently, she is part of the Human Cell Atlas consortium, with a focus on single-cell profiles and spatial characterization of gene expression in skeletal muscles at different developmental stages.

Chapter 1
The Role of Omics Approaches in Muscle Research

Stefano Schiaffino, Carlo Reggiani, and Marta Murgia

A thorough understanding of skeletal muscle physiology requires knowledge of the molecular composition of muscle tissue. So far, this has been mostly obtained through biochemical investigations focused on selected components of the muscle contractile and metabolic machinery. More recently, a systems biology approach, based on the emergence of omics technologies, has been introduced to get global views of all muscle components. These approaches have been used to explore the DNA (genomics and epigenomics), RNA (transcriptomics and non-coding RNA analyses), proteins (proteomics and global analyses of post-translational protein modifications, e.g. phosphoproteomics) and small molecules (metabolomics).

Genome-wide association studies (GWAS) compare single-nucleotide variants (single-nucleotide polymorphisms or SNPs) in the sequence of the whole human genome, to identify genes associated with muscle disorders that may influence the risk of the disease or modify the disease phenotype (genetic modifiers). Epigenetic changes, including both DNA methylation and histone post-translational modifications, have been analysed in muscle tissues to determine the effect of ageing, exercise and diet on gene expression. Microarray analyses and more recently RNA-seq techniques based on next-generation sequencing (NGS) provide a complete view of all mRNAs, as well as non-coding RNAs, including micro-RNAs and long non-coding RNAs. Transcriptome analyses have been widely applied to the

S. Schiaffino (✉)
Venetian Institute of Molecular Medicine (VIMM), Padova, Italy
e-mail: stefano.schiaffino@unipd.it

C. Reggiani
Department of Biomedical Sciences, University of Padova, Padova, Italy
e-mail: carlo.reggiani@unipd.it

M. Murgia
Department of Biomedical Sciences, University of Padova, Padova, Italy

Max-Planck-Institute of Biochemistry, Martinsried, Germany
e-mail: marta.murgia@unipd.it

J. G. Burniston, Y.-W. Chen (eds.), *Omics Approaches to Understanding Muscle Biology*, Methods in Physiology, https://doi.org/10.1007/978-1-4939-9802-9_1

study of skeletal muscle; however, not all transcripts are actually translated, and the analysis of proteins through proteomic approaches is required to identify the translated products in muscle cells. Proteomics and metabolomics, which are mostly based on liquid chromatography-mass spectrometry techniques, as well as nuclear magnetic resonance spectroscopy, have seen tremendous technological advances in the last years with both respect to instrumentation and data processing. Common proteomic and metabolomic experiments involve non-targeted or discovery approaches, which allow for the relative quantification of different proteins or metabolites, and targeted approaches, which allow for the absolute quantification of few selected components. A dynamic view of protein turnover, as a result of protein synthesis and degradation, can be obtained by combining proteomic analysis with protein labelling in vivo using isotope-labelled amino acids or heavy water [1].

Epigenetic changes have been extensively investigated in skeletal muscle. DNA methylation, leading to repression of gene transcription by reducing accessibility to gene promoters, is mediated by DNA methyltransferases (DNMTs) and is inhibited by specific demethylases, which usually activate transcription. DNA methylation involves transformation of cytosine into methyl-cytosine and can be quantified using bisulphite sequencing, taking advantage of the property of sodium bisulphite to convert unmethylated cytosine into uracil. Post-translational histone modifications affect chromatin condensation resulting in gene activation or gene silencing. They consist of acetylation, methylation, phosphorylation, ubiquitination and SUMOylation of specific amino acid residues of histones. Acetylation and deacetylation of histone lysine residues, which are especially important in regulating gene expression, are catalysed by histone acetyl transferases (HATs) and histone deacetylases (HDACs), respectively. Diet and exercise can modulate the muscle epigenome and even the possible transfer of epigenetic marks from one generation to the next [2].

Systematic analyses of alternative splicing, which produces distinct transcripts and protein isoforms for a large number of genes, have been combined with the study of specific functional interactions of each isoform. Widespread interaction differences due to alternative splicing were revealed from protein-protein interaction profiling of 1423 protein isoforms derived from alternative splicing of 506 human genes [3]. It will be of interest to apply similar interactome analyses to skeletal muscle proteins, many of which derive from alternative splicing. Indeed, skeletal muscle genes undergo extensive programs of alternative splicing, as shown in the Vertebrate Alternative Splicing and Transcription Database (VastDB) [4], although it remains to be established whether all of them are actually translated. This information is relevant to explain how mutations in widely expressed proteins can produce distinct pathologies in different tissues, including muscle, as shown in a recent study of tissue-specific splice variants for nuclear envelope proteins [5]. A further layer of complexity in the muscle proteome is generated by extremely high number and variety of post-translational modifications that can be measured in a cell. Advanced phosphoproteomic techniques allow to map more than 50,000 phosphorylation events on at least 75% of the proteome in a human cell line [6]. Acute high-intensity exercise was found to induce significant changes in human skeletal muscle phosphoproteome, including over 900 exercise-regulated phosphorylation sites that, as yet, have no known upstream kinase [7]. A novel proteomic approach, top-down mass spectrometry, allows to identify protein isoforms and post-translational modifications by analysing intact proteins without digestion: this approach has been used to characterize sarcomeric proteins from rhesus macaque skeletal muscle [8].

An integrated view of tissue metabolism can be obtained by combining different omics techniques, e.g. proteomics and metabolomics. This is especially useful when one tries to capture a temporal picture of metabolic changes, such as those taking place during the circadian cycle. This approach has been used to define the role of the intrinsic muscle clock in regulating the changes in glucose, lipid and amino acid metabolism during the feeding-fasting and rest-activity cycles [9, 10]. Lipid oscillations during the day/night cycle were also demonstrated by lipidomics analyses in human skeletal muscle and were found to persist in myotubes cultured in vitro [11]. Metabolic flux analyses using isotope tracers will be required to complement the snapshots provided by metabolomics studies and build a dynamic picture of metabolism. Finally, muscle metabolism must be considered in the context of the whole body, where complex crosstalk takes place between organs and tissues, as suggested by a global metabolomics analysis of circadian changes in eight mouse tissues, including skeletal muscle [12].

When omics approaches are applied to muscle samples, for example, human biopsies, it should be kept in mind that each sample contains a variety of other cell types. In a classic electron microscopy study, not always considered in omics studies, Schmalbruch and Hellhammer [13] reported that half of the nuclei in adult rat skeletal muscles are interstitial cell nuclei, predominantly fibroblasts and endothelial cells. This can obviously affect the interpretation of omics studies, for example, epigenomic analyses on muscle samples are expected to reflect in equal measure profiles of DNA methylation or histone post-translational modifications in muscle and non-muscle nuclei. Although myofibres certainly account for a greater proportion of the total proteins because of their large volume, the direct analysis of single myofibres can provide a more precise protein profile without any contamination by other cell types. A sensitive single-muscle-fibre proteomic technique has been developed to study mouse and human skeletal muscle [14, 15]. A crucial advantage of this approach is the possibility to compare for the first time the protein profile of the different fibre types present in a muscle biopsy and define their changes in different conditions. For example, single-muscle-fibre proteomics has shown that glycolytic enzymes decrease in fast 2A but increase in type1/slow muscle fibres during ageing [15]. Distinct protein changes induced by denervation in fast and slow muscle fibres were likewise revealed by single-muscle-fibre proteomics [16].

The availability of a quantitative portrait of muscle fibre proteomes will provide an essential foundation to interpret muscle fibre metabolism and physiology. A point to be stressed here is that the introduction of omics strategies does not simply result in the accumulation of a huge amount of new data. Instead, it can lead to conceptual advances by showing previously hidden correlations and enable researchers to frame new models and hypotheses. For example, single-muscle-fibre proteomics has shown that the NADP-dependent isocitrate dehydrogenase 2 (IDH2) is more abundant in type 1/slow mouse muscle fibres, whereas the NAD-dependent IDH3 is more abundant in fast 2X and 2B fibres, thus revealing the existence of alternative pathways in the Krebs cycle [14, 17]. The fibre-type-specific distribution of the two isoforms was confirmed by immunohistochemical staining with specific antibodies. The differential distribution of the two isoenzymes does not simply reflect the mitochondrial content, which is actually higher in type 2X compared to type 1 fibres. A plausible interpretation is that the continuously active slow fibres generate higher levels of reactive oxygen species (ROS), thus requiring more active antioxidant systems, which depend on sustained NADPH provision. If this interpretation is correct, one would expect that the other major NADPH-generating enzyme present in mitochondria, nicotinamide nucleotide transhydrogenase (NNT), which transforms NADH into NADPH, is likewise more abundant in slow fibre mitochondria. This was immediately verified by simple inspection of the proteomics database [17], thus supporting the hypothesis suggested by proteomic analyses, which can now be tested by appropriate knockdown experiments.

In conclusion, the rapid collection and examination of data by high-throughput omics techniques are useful not only to produce large datasets and identify biomarkers but also to generate new questions, models and hypotheses and has thus become an essential tool in muscle research. A crucial advantage of these studies compared to traditional studies based on candidate genes and proteins is that "data-driven" models and hypotheses are generated through an unbiased approach and are thus more likely to reveal new, often unexpected findings, whereas "hypothesis-driven" research is biased, thus more likely to confirm previous notions and findings [18, 19]. Obviously, the new leads suggested by the examination of large datasets must be followed by focused analyses aimed at verifying the new models. Indeed, the intelligent combination of data-driven and hypothesis-driven approaches, whereby data-driven predictions are validated by hypothesis-driven strategies, is going to become a standard approach to science with the best chances to increase our knowledge in all fields of biology, including muscle biology.

References

1. Camera, D. M., Burniston, J. G., Pogson, M. A., Smiles, W. J., & Hawley, J. A. (2017). Dynamic proteome profiling of individual proteins in human skeletal muscle after a high-fat diet and resistance exercise. *The FASEB Journal, 31*, 5478–5494.
2. Barrès, R., & Zierath, J. R. (2016). The role of diet and exercise in the transgenerational epigenetic landscape of T2DM. *Nature Reviews Endocrinology, 12*, 441–451.
3. Yang, X., Coulombe-Huntington, J., Kang, S., Sheynkman, G. M., Hao, T., Richardson, A., Sun, S., Yang, F., Shen, Y. A., Murray, R. R., Spirohn, K., Begg, B. E., Duran-Frigola, M., MacWilliams, A., Pevzner, S. J., Zhong, Q., Trigg, S. A., Tam, S., Ghamsari, L., Sahni, N., Yi, S., Rodriguez, M. D., Balcha, D., Tan, G., Costanzo, M., Andrews, B., Boone, C., Zhou, X. J., Salehi-Ashtiani, K., Charloteaux, B., Chen, A. A., Calderwood, M. A., Aloy, P., Roth, F. P., Hill, D. E., Iakoucheva, L. M., Xia, Y., & Vidal, M. (2016). Widespread expansion of protein interaction capabilities by alternative splicing. *Cell, 164*, 805–817.

4. Tapial, J., Ha, K. C. H., Sterne-Weiler, T., Gohr, A., Braunschweig, U., Hermoso-Pulido, A., Quesnel-Vallières, M., Permanyer, J., Sodaei, R., Marquez, Y., Cozzuto, L., Wang, X., Gómez-Velázquez, M., Rayon, T., Manzanares, M., Ponomarenko, J., Blencowe, B. J., & Irimia, M. (2017). An atlas of alternative splicing profiles and functional associations reveals new regulatory programs and genes that simultaneously express multiple major isoforms. *Genome Research, 27*(10), 1759–1768.
5. Capitanchik, C., Dixon, C. R., Swanson, S. K., Florens, L., Kerr, A. R. W., & Schirmer, E. C. (2018). Analysis of RNA-Seq datasets reveals enrichment of tissue-specific splice variants for nuclear envelope proteins. *Nucleus, 9*, 410–430.
6. Sharma, K., D'Souza, R. C., Tyanova, S., Schaab, C., Wiśniewski, J. R., Cox, J., & Mann, M. (2014). Ultradeep human phosphoproteome reveals a distinct regulatory nature of Tyr and Ser/Thr-based signaling. *Cell Reports, 8*, 1583–1594.
7. Hoffman, N. J., Parker, B. L., Chaudhuri, R., Fisher-Wellman, K. H., Kleinert, M., Humphrey, S. J., Yang, P., Holliday, M., Trefely, S., Fazakerley, D. J., Stöckli, J., Burchfield, J. G., Jensen, T. E., Jothi, R., Kiens, B., Wojtaszewski, J. F., Richter, E. A., & James, D. E. (2015). Global phosphoproteomic analysis of human skeletal muscle reveals a network of exercise-regulated kinases and AMPK substrates. *Cell Metabolism, 22*, 922–935.
8. Jin, Y., Diffee, G. M., Colman, R. J., Anderson, R. M., & Ge, Y. (2019). Top-down mass spectrometry of sarcomeric protein post-translational modifications from non-human primate skeletal muscle. *Journal of the American Society for Mass Spectrometry*. https://doi.org/10.1007/s13361-019-02139-0.
9. Dyar, K. A., Ciciliot, S., Wright, L. E., Bienso, R. S., Tagliazucchi, G. M., Patel, V. R., Forcato, M., Paz, M. I., Gudiksen, A., Solagna, F., Albiero, M., Moretti, I., Eckel-Mahan, K. L., Baldi, P., Sassone-Corsi, P., Rizzuto, R., Bicciato, S., Pilegaard, H., Blaauw, B., & Schiaffino, S. (2014). Muscle insulin sensitivity and glucose metabolism are controlled by the intrinsic muscle clock. *Molecular Metabolism, 3*, 29–41.
10. Dyar, K. A., Hubert, M. J., Mir, A. A., Ciciliot, S., Lutter, D., Greulich, F., Quagliarini, F., Kleinert, M., Fischer, K., Eichmann, T. O., Wright, L. E., Pena Paz, M. I., Casarin, A., Pertegato, V., Romanello, V., Albiero, M., Mazzucco, S., Rizzuto, R., Salviati, L., Biolo, G., Blaauw, B., Schiaffino, S., & Uhlenhaut, N. H. (2018). Transcriptional programming of lipid and amino acid metabolism by the skeletal muscle circadian clock. *PLoS Biology, 16*, e2005886.
11. Loizides-Mangold, U., Perrin, L., Vandereycken, B., Betts, J. A., Walhin, J. P., Templeman, I., Chanon, S., Weger, B. D., Durand, C., Robert, M., Paz Montoya, J., Moniatte, M., Karagounis, L. G., Johnston, J. D., Gachon, F., Lefai, E., Riezman, H., & Dibner, C. (2017). Lipidomics reveals diurnal lipid oscillations in human skeletal muscle persisting in cellular myotubes cultured in vitro. *Proceedings of the National Academy of Sciences of the United States of America, 114*, E8565–E8574.
12. Dyar, K. A., Lutter, D., Artati, A., Ceglia, N. J., Liu, Y., Armenta, D., Jastroch, M., Schneider, S., de Mateo, S., Cervantes, M., Abbondante, S., Tognini, P., Orozco-Solis, R., Kinouchi, K., Wang, C., Swerdloff, R., Nadeef, S., Masri, S., Magistretti, P., Orlando, V., Borrelli, E., Uhlenhaut, N. H., Baldi, P., Adamski, J., Tschop, M. H., Eckel-Mahan, K., & Sassone-Corsi, P. (2018). Atlas of circadian metabolism reveals system-wide coordination and communication between clocks. *Cell, 174*, 1571–1585.e1511.
13. Schmalbruch, H., & Hellhammer, U. (1977). The number of nuclei in adult rat muscles with special reference to satellite cells. *The Anatomical Record, 189*, 169–175.
14. Murgia, M., Nagaraj, N., Deshmukh, A. S., Zeiler, M., Cancellara, P., Moretti, I., Reggiani, C., Schiaffino, S., & Mann, M. (2015). Single muscle fiber proteomics reveals unexpected mitochondrial specialization. *EMBO Reports, 16*, 387–395.
15. Murgia, M., Toniolo, L., Nagaraj, N., Ciciliot, S., Vindigni, V., Schiaffino, S., Reggiani, C., & Mann, M. (2017). Single muscle fiber proteomics reveals fiber-type-specific features of human muscle aging. *Cell Reports, 19*, 2396–2409.

16. Lang, F., Khaghani, S., Türk, C., Wiederstein, J. L., Hölper, S., Piller, T., Nogara, L., Blaauw, B., Günther, S., Müller, S., Braun, T., & Krüger, M. (2018). Single muscle fiber proteomics reveals distinct protein changes in slow and fast fibers during muscle atrophy. *Journal of Proteome Research, 17*, 3333–3347.
17. Schiaffino, S., Reggiani, C., Kostrominova, T. Y., Mann, M., & Murgia, M. (2015). Mitochondrial specialization revealed by single muscle fiber proteomics: Focus on the Krebs cycle. *Scandinavian Journal of Medicine and Science in Sports, 25*(Suppl 4), 41–48.
18. Glass, D. J. (2010). A critique of the hypothesis, and a defense of the question, as a framework for experimentation. *Clinical Chemistry, 56*, 1080–1085.
19. Van Helden, P. (2013). Data-driven hypotheses. *EMBO Reports, 14*, 104.

Part I
Genomic and Epi-genomic

Chapter 2
Genome-Wide Association Studies in Muscle Physiology and Disease

Luca Bello, Elena Pegoraro, and Eric P. Hoffman

2.1 Historical Outline and Rationale

The ultimate goal of genetic studies is to identify the role of genes in determining health and disease traits, in a complex context where factors distinct from genetics (i.e., environmental) may also alter these traits. The concept of "heritability" defines in what amount the variance of a specific trait in a population is dictated by inherited genes, as opposed to the environment. From this perspective, genetics may be viewed as the study of heritability and its molecular underpinnings.

Given the endless diversity of physiological and pathological traits that may be taken in consideration in humans and other organisms, it is intuitive that the study of their heritability may be approached from many different angles and with many different research tools. A distinction could be made between the study of common DNA variations (usually called "polymorphisms") seeking to explain health traits or risk of common diseases (e.g., muscle mass or risk of diabetes), as opposed to rare DNA variants (usually called "mutations") that cause Mendelian diseases (e.g., Duchenne muscular dystrophy or cystic fibrosis) with a high penetrance. The genome-wide association study (GWAS) essentially addresses the former scenario, and it may be considered the main research tool used to approach common genetic variation at the genomic level. As a means to map common genetic variation, the GWAS takes advantage of the most abundant form or variation in our genomes: single-nucleotide polymorphisms (SNPs).

SNPs are not only the most abundant but also the simplest form of genetic variant. For the most part, as suggested by the name, SNPs consist of the substitution of a

L. Bello (✉) · E. Pegoraro
Department of Neurosciences DNS, University of Padova, Padova, Italy
e-mail: luca.bello@unipd.it

E. P. Hoffman
School of Pharmacy and Pharmaceutical Sciences, Binghamton University, Binghamton, NY, USA

J. G. Burniston, Y.-W. Chen (eds.), *Omics Approaches to Understanding Muscle Biology*, Methods in Physiology, https://doi.org/10.1007/978-1-4939-9802-9_2

single nucleotide in the DNA sequence with another nucleotide: purine-purine and pyrimidine-pyrimidine changes are called transitions, while purine-pyrimidine and pyrimidine-purine changes are called transversions. However, more complex changes such as small rearrangements (e.g., deletions-insertions of a few nucleotides), as well as polymorphisms with more than two possible alleles (multiallelic vs. biallelic), may be included in an extended definition of an SNP. Importantly, mutations that cause rare diseases do not at all differ structurally from common SNPs at the DNA level, but are just rarer, and have more deleterious effects on RNA and/or protein expression and function.

Since the publication of the draft sequence of the human genome in 2001 [1, 2], it became evident that SNPs are present in approximately 1 in 1300 position of the human genetic sequence, around 1.42 million SNPs having been identified by the International Human Genome Sequencing Consortium [3]. SNPs arise from random mutational events and, if not purged by natural selection, as may be the case with rare disease-causing mutations, are then inherited in subsequent generations. The frequency of a certain SNP in the population is usually defined by its minor allele frequency (MAF). If, for instance, 70% of chromosomes in a population show a cytosine base at a certain position, while 30% of chromosomes show a transition to a thymine base, an SNP with a 0.30 MAF is determined to be present in that population.

Most of the time, SNPs have little or no effect in themselves on the expression or function of neighboring genes but are simply co-inherited with alleles that are associated with true biological effects because of linkage disequilibrium (LD), i.e., the low frequency of recombination events happening between DNA stretches that are physically close together. Because of the LD structure of genomes, SNPs may be used as "tags" for specific alleles at genomic loci in which they are situated.

Since the early 2000s, the development of relatively inexpensive SNP genotyping arrays, allowing to genotype multiple individuals at up to 10^5–10^6 SNPs throughout the genome in a single experiment, provided an easily accessible means of mapping genes that influence complex phenotypic traits and disease risks. It should be again stressed that SNPs genotyped by such arrays seldom have a direct influence on the trait or disease of interest but may be in LD with functional variants in neighboring genes, providing a rationale for using them as tags for quantitative trait loci (QTLs) or disease-risk alleles.

Operatively, GWAS is based on massively parallel statistical tests of association between the phenotype of interest and genotype at each genotyped SNP. For instance, in a classic case-control design for disease risk, MAFs of genotyped SNPs are compared between affected and control individuals, and a χ^2 statistic and corresponding odds ratio (OR), i.e., an estimation of risk associated to carrying a copy of the minor allele, is calculated for each SNP. An OR > 1 means an increased risk of disease, while OR < 1 means that the minor allele is protective. Alternatively, in a linear association design usually adopted for the study of QTLs, the quantitative phenotype is regressed against the genotype modeled as a quantitative variable (e.g., 0 for common allele homozygotes, 1 for heterozygotes, and 2 for minor allele homozygotes in additive models), calculating a linear correlation coefficient for

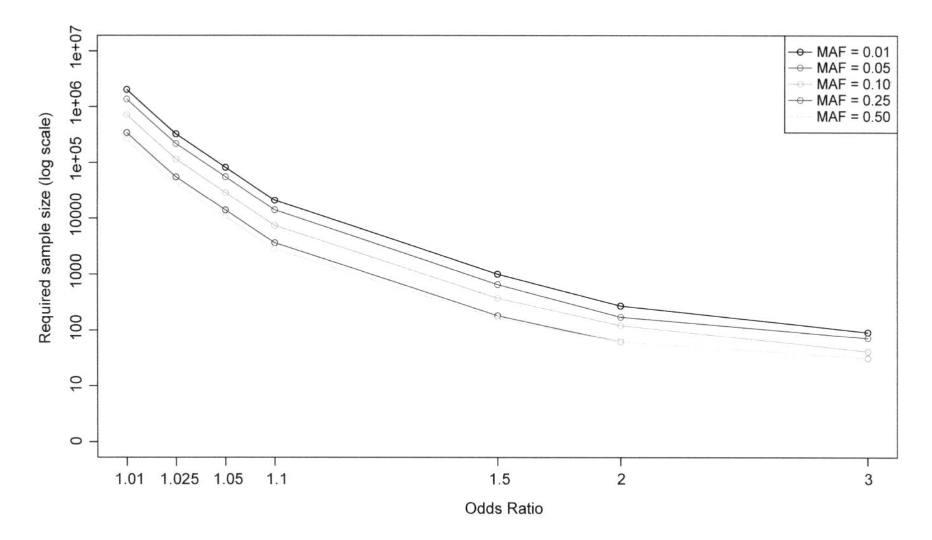

Fig. 2.1 Power calculation showing required GWAS sample size (number of individuals tested) as a function of odds ratio (OR; estimated disease risk per copy of the risk allele) for SNPs with different minor allele frequencies (MAFs), spanning the MAF spectrum usually included in genotyping arrays [MAF > 0.01 (1% allele frequency)]. Note that the *y*-axis (sample size) scale is logarithmic, while the *x*-axis (OR) is differently scaled for small ORs (<1.1) for better clarity

each SNP. Because of the massively parallel nature of the tests, in which allele frequency at hundreds of thousands of loci is compared between the same two groups of individuals, or the same set of quantitative data is regressed against genotypes at hundreds of thousands of loci, it is intuitive that a strict correction for multiple testing must be enforced, leading to a very low *p*-value threshold for establishing statistical significance of genome-wide associations. This threshold is usually set at 5×10^{-8} in GWASs involving individuals of European ancestry, assuming around a million independently inherited common variants [4, 5], and thus dividing the usual 0.05 threshold of statistical significance by a million as a Bonferroni correction for the number of parallel tests. In order to attain sufficient statistical power in these conditions (e.g., predicted reaching the 5×10^{-8} *p*-value threshold), it is usually necessary to sample numerous individuals, starting from several hundred up to tens of thousands, depending mainly on the strength of the genotype-phenotype association and on the frequency of the associated variant. A power calculation illustrating required sample sizes with different values of these variables is shown in Fig. 2.1.

It appears clearly from Fig. 2.1 that the required sample for identifying a statistically significant genotype-phenotype association by GWAS decreases as a function of the effect size of the associated variant (odds ratio; OR). For instance, several thousands to tens of thousands of individuals are needed to identify a risk variant with a small/moderate OR of 1.1, while a smaller sample size would be sufficient for larger effects, e.g., a few hundred individuals for ORs over 2. Very large GWASs or meta-analyses of multiple GWASs, with collective sample sizes reaching hundreds of thousands of genotyped participants, are needed to pinpoint SNP associations with small effect sizes, in the order of 1.01–1.05 (e.g., 1–5% increased risk). Both with high and low ORs, it is easier to identify the effect of common (MAF above 0.05) rather than infrequent (MAF 0.01–0.05) variants. In essence, the statistics of GWAS are similar to any other experiment: the bigger the effect of a variable, the easier it is to prove it is not a false positive. However, GWASs, unlike other experiments, have to compensate for the very large number of tests being done (thousands or millions of genotypes run against a phenotype), and this leads to a much higher p-value threshold for significance ($p < 5 \times 10^{-8}$).

"Classic" GWAS algorithms cannot at all be applied to rare variants (MAF below 0.01), identified by next-generation sequencing (NGS) techniques such as whole exome sequencing (WES) or whole genome sequencing (WGS), as the number of participants carrying the minor allele would be too small to sufficiently power a study. In order to circumvent this problem, specific association algorithms for rare variants have been devised, which will be discussed separately. Typically, genetic associations identified by successful GWASs have shown small-to-moderate size effects (OR 1.05–1.5), with sample sizes in the order of 10^3 (1000 subjects studied).

In fact, because of the purging effect of natural selection, it is unlikely for common variants targeted by GWAS arrays to have large effects on disease risks or common traits. There are, however, exceptions to this rule, especially for diseases that do not alter reproductive fitness. For example, the very first published GWAS [6] identified a true risk variant for age-related macular degeneration within the *CFH* gene, by genotyping only 96 cases and 50 controls. The associated OR was very high (7.4). The following years of GWAS research, on the other hand, have revealed that such "low-hanging fruits" are extremely unusual.

Figure 2.2 exemplifies different kinds of genetic associations according to the effect size/allele frequency ratio. Variants with large effect size and common allele frequency (top right of the diagram) correspond to the "low-hanging fruits" mentioned above. On the bottom right, with small-to-intermediate effect size and common allele frequency, are associations usually identified by typical GWASs. Rare variants with strong effects (top left) correspond to the disease-causing mutations of Mendelian diseases, which are not captured by genotyping chips and are best identified by targeted sequencing or NGS in individuals or families with suggestive phenotypes. On the other hand, the field of rare variant association in common diseases and traits is an emerging one, which requires specific algorithms different from "classic" GWAS (discussed separately). Finally, rare variants with small effects (bottom left) may be hard or impossible to identify. While the individual role of any such variant in disease and physiology may be of secondary importance, the collective role of large numbers of variants with small effects is very challenging to gauge and represents a potential source of "missing heritability" [7]. Other potential sources of "missing heritability" are types of genetics not measured by GWAS (e.g., epigenomic effects, or very rare mutations with very large effects).

Since the first successful attempt in 2005, the GWAS approach has led to both impressive genetic associations and failures (major factors in traits such as obesity or fitness that are not queried by GWAS—e.g., diet and effort, respectively). The

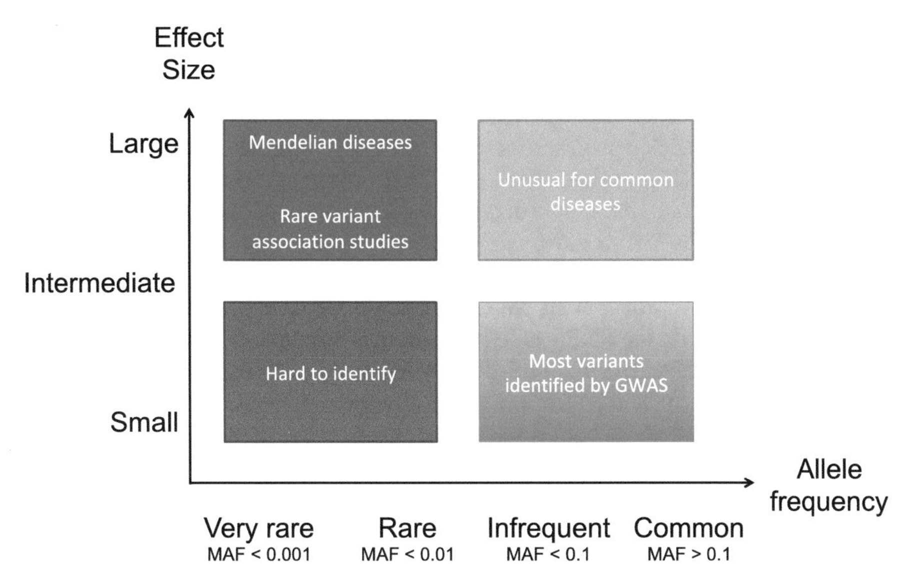

Fig. 2.2 Diagram of the size effect/allele frequency ratio of genetic associations and corresponding approaches leading to their identification. Modified from [5]

National Human Genome Research Institute (NHGRI), in collaboration with the European Molecular Biology Laboratory-European Bioinformatics Institute (EMBL-EBI), currently lists 68,147 different associations from 2847 published studies (Hindorff LA, MacArthur J [EBI], Morales J [EBI], Junkins HA, Hall PN, Klemm AK, and Manolio TA. A Catalog of Published Genome-Wide Association Studies. Available at: www.genome.gov/gwastudies. Accessed April 25th, 2018).

These myriad studies have greatly increased our knowledge on the genetic architecture and molecular mechanisms underlying common disorders and traits, both in general and specifically for individual phenotypes. In general, it has been established that virtually all common diseases and traits are polygenic, i.e., they are associated with several independent loci in different chromosomic regions. Conversely, there is pervasive pleiotropy, i.e., individual variants are associated with several different phenotypes, a classic example being variants that represent risk factors in some autoimmune diseases and protective factors for others. Several novel disease pathways and therapeutic targets have been identified by GWAS, and a recent excellent review can be found here [8]. This chapter will be aimed at application of GWAS to muscle diseases and muscle physiology and focus on methodological aspects.

2.2 Outline of GWAS Methods

Genotyping Assays and Genotype Calling Genotyping assays in SNP genotyping chips are based on allele-specific oligonucleotide probes complementary to a known SNP-containing DNA sequence. The target DNA is fragmented and hybridized with the probes, which are immobilized on the solid surface of the chip at a known position for each SNP. The most successful and widely used commercial designs were those by Affymetrix, based on 25-mer probes carrying the SNP locus at varying positions within the probe, with which the target DNA hybridizes more or less strongly depending on the SNP genotype, leading to stronger or dimmer fluorescent signal, and by Illumina, based on the 50-mer sequence immediately adjacent to the SNP and fluorescently labeled free base extensions that bind specifically to different alleles in the target DNA. In both cases, different fluorescent signals are produced for each allele, allowing to discriminate major allele homozygotes (presence of fluorescent signal corresponding to the major allele only), heterozygotes (presence of fluorescent signals corresponding to both alleles), and minor allele homozygotes (presence of fluorescent signal corresponding to the minor allele only). After hybridization, fluorescent signals are read at each position of the array by automated fluorescence microscopy, and genotypes are called at each SNP position from raw fluorescence intensity data, using dedicated software. Genotype calling for biallelic SNPs (which constitute the vast majority of assayed SNPs) is based on "clustering" of samples according to fluorescence intensity (example in Fig. 2.3a). Whenever samples are not closely "clustered" within areas defined by the genotype calling

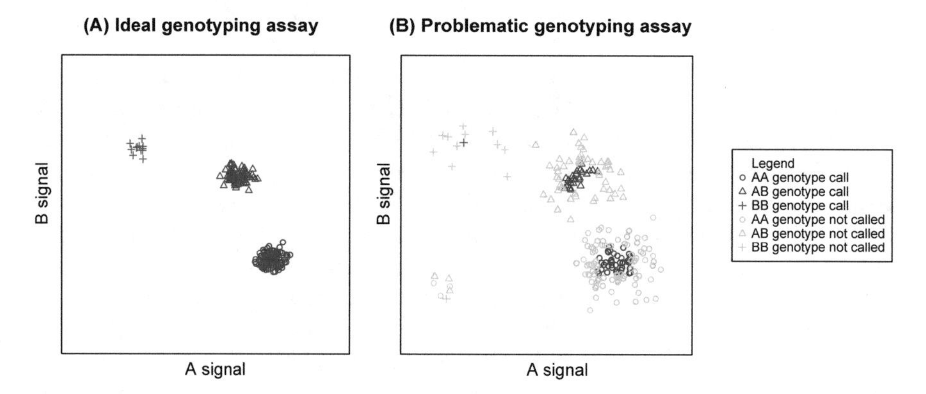

Fig. 2.3 Examples of genotyping assay results for a biallelic SNP, plotted with fluorescent signal corresponding to the "A" allele (major allele) on the *x*-axis and fluorescent signal corresponding to the "B" (minor) allele on the *y*-axis. In the (**a**) panel, an ideal assay is represented, where samples closely cluster according to genotype. In the (**b**) panel a problematic assay is represented, where fluorescence intensity values are variable across samples with the same genotype, so that several samples are distant from the cluster center. In a few samples (bottom left), no signal is observed because of experimental failure. The result is a high percentage of missed calls (gray data points) that may be reduced by increasing the area of called genotype clusters. However, this would imply a risk of wrong calls in borderline areas

algorithm (which may be more strict or permissive depending on experimental conditions), the results are missing genotype calls (Fig. 2.3b). A relevant percentage of missing calls in an SNP assay across samples, or in a sample across assays, may indicate faulty assays or bad-quality DNA samples, respectively.

Data Quality Checks and Data Cleaning Once genotype at all SNP positions has been called, before genome-wide genotype data may be used in association algorithms, strict quality checks and data cleaning procedures are essential. Including error-ridden data in the analyses exposes researchers to a risk of false-positive and false-negative findings. Common errors and sources of bias in GWASs include wrong genotypes, sample mislabeling, cryptic duplicates, wrong phenotype data, sample sexing errors, and wrong association of samples with phenotype data. A peculiar source of bias, population stratification, is often caused by cryptic relatedness and by the compresence of subgroups with heterogeneous ancestry within the GWAS sample and will be discussed separately.

Missingness As illustrated in Fig. 2.3, issues with genotyping assays may lead to missing genotype calls. A faulty genotyping assay, illustrated in panel (b), will yield a high percentage of missed calls across samples, and DNA samples of insufficient quality or quantity will yield a high percentage of missed calls across assays, as exemplified by the small cluster of samples with very low signals for both alleles in panel (b). A threshold for missingness is often the first step in GWAS data cleaning, mainly because even non-missing data that derive from assays and samples with high missingness are probably error-ridden and would lead to false-positive or false-negative findings. The specific thresholds used for filtering vary according to experimental design and conditions, but common examples are removal of samples/assays with >5% missingness. Assays for SNPs with low MAF (below 5%) require a more stringent threshold (e.g., 1%).

Relatedness and Duplicates From genotype calls in the whole sample, it is possible to compute an "identity-by-state" (IBS) matrix, containing information about how similar/dissimilar any couple of samples in the genotyped population is. IBS expresses the concept of identical genotype at any particular SNP position, not necessarily implying "identity by descent" (IBD), which means that a certain allele has been inherited from a common ancestor in two or more individuals. In order to quantify the degree of genome-wide IBS between two samples, the "pi-hat" parameter is often used, where "1" identifies identical twins or the same sample genotyped twice, 0.5 first-degree relatives like siblings or parent-offspring pairs, 0.25 second-degree relatives, and so on. Ideally, GWASs reach the greatest statistical power when participants are of homogeneous ancestry but are not closely related with each other. This guarantees that SNPs tag the same true functional variants because of homogeneous haplotype structure within the population, at the same time minimizing population substructure bias (see below). Even if study recruitment is designed in such a way as to avoid relatedness, a genotype-level verification can be very helpful. For instance, especially in large multicenter studies, the same individual might be enrolled in the same study twice (knowingly or unknowingly), or related individuals might be inadvertently enrolled. Alternatively, samples might be duplicated because of laboratory handling errors. An IBS matrix is an easy way to identify and remove cryptic duplicates and related samples.

Heterozygosity The degree of heterozygosity for common SNP genotype is variable between individuals and between genomic regions. Higher-than-usual degrees of genome-wide heterozygosity in individual samples might indicate recent genetic admixture, but in the context of a GWAS experiment, it might simply signify sample contamination. Therefore, an inspection of heterozygosity levels among the GWAS population and elimination of outlier samples (e.g., >4 standard deviations from the mean) is advisable.

Sex Sex of recruited individuals is usually annotated in the phenotypic information. Sex may be relevant in GWAS both for associations involving X chromosome loci and for its influence as an "environmental" modifier of phenotype (e.g., hormonal influences, sexual dimorphisms, behavioral issues, etc.). Based on X chromosome SNP genotyping, it is often possible to determine an individual's sex from SNP chip data, although multiple SNP loci need to be scored for accuracy. A comparison of phenotype and SNP-derived sex data is advisable during SNP chip data cleaning, in order to double-check discordant samples.

Hardy-Weinberg Equilibrium (HWE) Usually, common SNPs included in GWAS chips are polymorphisms that have stable allele frequencies in a population as there is no evolutionary pressure and because unrelated individuals of homogeneous ancestry are recruited into a typical GWAS. Therefore, it is expected that observed genotypes do not violate HWE, as expressed by the formula $p^2 + 2pq + q^2 = 1$, where p is the major allele frequency and q the minor allele frequency. A χ^2 test for deviation from HWE at each genotyped SNP can be performed to verify this assumption. Deviations from HWE might be caused by population substructure or genotyping errors. However, because of the sheer amount of genotyped SNPs in a GWAS, HWE violations with $p < 0.05$ are always observed, and removal of SNPs from the GWAS workflow because of HWE violation is not recommended. SNPs with lower HWE p-values (e.g., $p < 10^{-5}$) may be "flagged" for technical validation of genotyping, in case of a positive association with phenotype. Importantly, in case-control studies, a HWE violation in cases may actually be caused by a positive association with the disease, as the sampling of these individuals from the population is not random. Therefore, HWE should be calculated separately for cases and controls.

Population Stratification The concept of population stratification identifies a situation in which the GWAS sample is subdivided in clusters that systematically differ in allele frequency at some loci, most commonly because of differences in ancestry. This may be because of relatedness (overt or cryptic) between individuals in the sample, because of heterogeneous racial/ethnic groups within the sample. Population stratification is a major source of bias in GWAS, mainly because individuals who are related, or belong to similar racial/ethnic groups, may easily be subject to similar environmental or other nongenetic factors that modify traits or disease risks over and above the genetics. Therefore, all variants that have a different MAF in that group will appear associated with the phenotype, leading to false-positive results (inflation of type I error). Secondarily, as haplotype structure differs between populations, the same SNP may function as an efficient tag of a certain risk allele in one population, but not in another, resulting in loss of power and false-negative results at some loci (inflation of type II error). Therefore, as mentioned earlier, "classic" GWASs are aimed at recruiting individuals who are unrelated but identify within the same racial/ethnic group. As a result of striving towards homogeneity of the racial/ethnic subjects to increase accuracy of GWASs, the design of the most widely used SNP genotyping chips is largely based on the haplotype structure of populations of European descent. However, with growing knowledge about the haplotype structures of diverse human populations, due to the success of sequencing projects such as the 1000 Genomes project, SNP chips are increasingly becoming available, which are adapted to the study of population with different and/or more complex and variable haplotype structures, such as populations of African descent.

A simple, effective way of inspecting genotype data for the presence of population stratification is analysis of principal components (PCs). PCs are mathematical vectors that describe major patterns of similarity/dissimilarity among samples. Whenever samples tend to cluster in different groups according to values of one or more PCs, it can be concluded that population stratification is present. Figure 2.4 exemplifies this concept by illustrating a plot of the two main PCs from multidimensional scaling (MDS, an algorithm akin to PC analysis) analysis of genome-wide genotype data from the Cooperative International Neuromuscular Research Group Duchenne Natural History Study (CINRG-DNHS) [9]. As the CINRG-DNHS was an international study recruiting a diverse, multiethnic cohort across four continents, it appears clearly from the MDS plot that there is substantial population stratification, as a main cluster of participants are concentrated on the left of the plot (low values of the first PC), while a consistent minority of samples tended to "drift" rightwards with higher values of the first PC and alternatively higher or lower values of a second PC. From racial/ethnic self-identification data (according to categories recognized by the United States Census Bureau), it is clear that higher values of both PC1 and PC2 are associated with participants of African descent (top right), while participants of Asian descent present higher values of PC1 and lower values of PC2 (bottom right), and the leftward cluster corresponds to a core of participants of European ancestry. Muscular dystrophy phenotype is associated to ethnicity because of nongenetic reasons, i.e., regional standards of care and cultural/economic factors; therefore, the simplest choice in order to minimize population stratification bias in such a scenario is to limit association analyses to the main cluster of participants of European descent (left of the arbitrary PC1 cutoff indicated by the red line in the plot). However, alternative approaches are possible such as clustered association algorithms or implementations of selected PCs as covariates.

Main Association Algorithms For case-control studies (the most typical approach for common disease GWASs), the association test that is run in parallel for all SNPs is a basic allelic χ^2 test comparing allele frequency between cases and controls (2×2

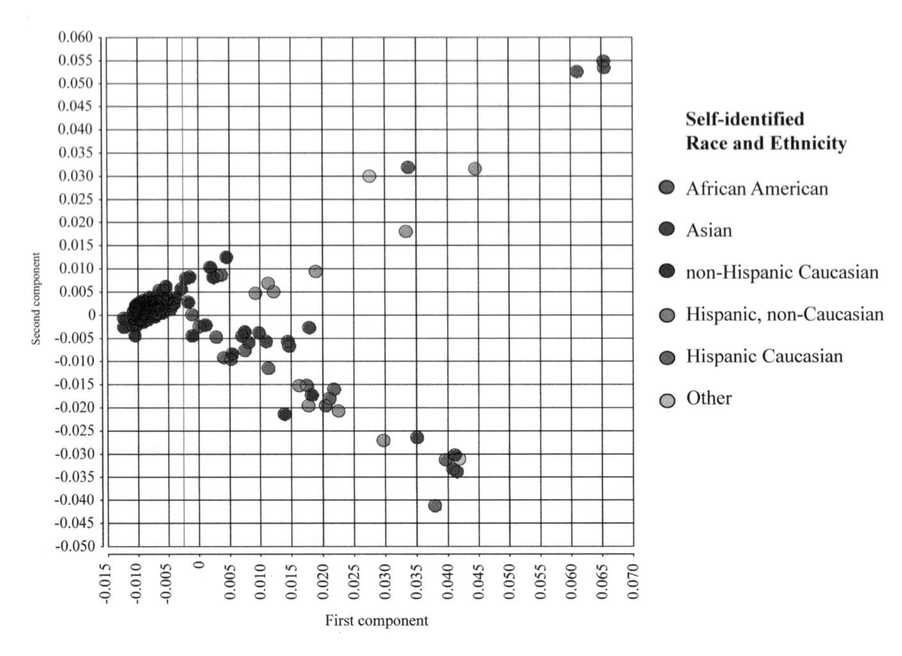

Fig. 2.4 Multidimensional scaling (MDS) analysis of genome-wide genotypes in the population of participants in the Cooperative International Neuromuscular Research Group Duchenne Natural History Study. Each data point represents an individual. Position in the Cartesian plot is dictated by the first two principal components deriving from MDS, while color indicates self-identified ancestry (regardless of genotype data). The vertical red line represents the arbitrary cutoff adopted in order to select a homogeneous subpopulation and minimize population stratification bias in association analyses

table with 1 degree of freedom) and calculating an estimate of risk per copy (additive model) of the minor allele (OR) and its corresponding p-value. When rare SNPs are included that might lead to values ≤ 5 in any cell of the 2×2 tables, use of Fisher's exact test is recommended (although this would also imply low statistical power of the GWAS). Dominant and recessive inheritance models are seldom used in GWAS, as fully dominant/recessive effects are more typical of disrupting, high-penetrance, Mendelian gene mutations than of more common variants with mild effects on common diseases and traits. Furthermore, even when a true functional variant does lean towards dominance/recessivity, it is likely that its tag SNP on the genotyping chip will yield the strongest signal through an additive model, because of non-perfect LD between the tag SNP and the functional variant. All these considered, dominant/recessive case-control studies do not differ much analytically from their additive counterpart, as the only modification needed is a subject-based, rather than allele-based χ^2 test. Genotypic, non-additive tests (e.g., 2×3 table with 2 degrees of freedom for AA, AB, and BB genotypes) are also rarely adopted. Quantitative trait (QT) association, on the other hand, is based on regression between the quantitative phenotype and a quantitative model of the genotype. As SNPs included in GWAS chips are typically biallelic, the most common genotype model is additive, with major allele homozygote coded as 0, heterozygote as 1, and minor allele homozygote as 2, which corresponds to number of copies of the minor allele. A regression coefficient β and corresponding p-value are calculated for each SNP. Different codes (e.g., 0-1-1 for dominant, 0-0-1 for recessive, 1-2-4 for multiplicative) allow exploration of different genotypic models, according to the underlying scientific hypothesis. Then again, GWASs are mostly hypothesis-free, discovery studies, and the balanced additive model is by far the most widely used. Statistical models based on logistic and linear regression allow the implementation of covariates for case-control and QT data, respectively. Covariates are often of the utmost importance in order to reach valid conclusions. For instance, major nongenetic risk factors, such as cigarette smoke or diet, account for large amounts of the variance of phenotypic outcomes. When this is controlled for by specific covariates, statistical power for the identification of genetic effects substantially rises. Time- and age-related covariates may be implemented through the use of time-to-event statistical models (e.g., Cox regression).

Quantile-Quantile (QQ) and Manhattan Plots An effective way to present overall GWAS results is plots of association p-values at each individual SNPs. The negative logarithm with base 10 of the p-value is plotted on the y-axis (e.g., $p = 0.00001$, $-\log10(p) = 5$), so that highly associated signals appear as high values in the plot. A QQ plot is used to inspect GWAS result for systematic bias (such as population stratification), plotting observed $-\log10(p\text{-values})$ against theoretical $-\log10(p\text{-values})$ that would be observed under the null hypothesis, i.e., in conditions of no association whatsoever between phenotypes and SNP genotypes. In the absence of systematic bias, all p-values will be distributed on a straight line corresponding to expected values under the null hypothesis, except for p-values corresponding with few truly associated SNPs, which will have higher observed values with genome-wide, or at least suggestive, significance. If there is an early deviation of observed value from expected values under the null hypothesis, some source of systematic bias must be suspected. Upward deviation in the QQ plot signifies type I error inflation (risk of false-positive results), as seen in the case of population stratification, while downward deviation indicates type II error inflation (risk of false-negative results). The Manhattan plot shares the same y-axis as the QQ plot, but on the x-axis, SNPs are arranged according to physical chromosome position. Association signals appear as groups of p-values "towering" above the others (like skyscrapers, hence the "Manhattan" name), with a highest point corresponding to the individual SNP or SNPs best tagging the hypothetical true functional variant, and a trail of SNPs showing lower association deriving from decreasing LD (Fig. 2.5).

The identification of a statistically significant association signal, in line with the hypothesis-generating, discovery-oriented nature of GWAS, represents the basis for experimental approaches aimed at characterizing and mechanistically defining the association. Validation in independent GWASs and meta-analyses of multiple GWASs, possibly across populations of different ancestry, are needed to confirm SNP association and definitely establish their effect size. Regarding functional characterization, first steps usually involve targeted high-density genotyping or

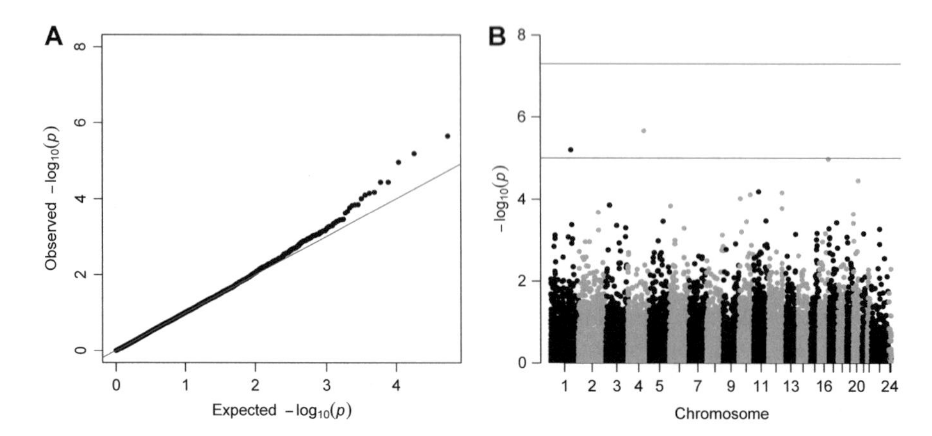

Fig. 2.5 QQ plot (**a**) and Manhattan plot for an exome-chip GWAS of age at loss of ambulation in 109 participants of European ancestry enrolled in the CINRG Duchenne Natural History Study [10]. (**a**) The QQ plot shows that observed *p*-values (black dots) are very similar to those expected under the null hypothesis (red line) up until the highest −log10 *p*-values, where a few *p*-values are higher than expected (putative association). (**b**) Manhattan plot of the same data plotted against chromosome position. Note that in this GWAS, no loci surpass the genome-wide significance threshold of 5×10^{-8} (red line in the Manhattan plot), with only a few suggestive associations (blue threshold line arbitrarily drawn at 10^{-5})

resequencing of nearby genes, especially in genomic areas that represent LD blocks in the population of interest, in order to possibly identify the true functional variant and its effect on corresponding genes. Subsequent functional studies vary based on the relevant biology.

2.3 Applications of GWAS to Muscle Biology

2.3.1 Applications in Muscle Physiology

Skeletal muscle is one of the largest organ systems of the body of all mammals, and also one of the most adaptable tissues, showing dramatic remodeling in response to different types of exercise and training, diet and fasting, and development and aging. Skeletal muscle is also a major industry in terms of "meat"—many animals have been genetically bred to increase meat volume and taste. Given the commercial importance of muscle as a food, it may not be unexpected that many GWASs have focused on muscle in terms of meat quantity and quality [11–13]. Due to the nature of strong pressures of human selection for animal traits, it is very rare for animal genetics to relate to human genetics (more "natural" selection in the latter).

In terms of human GWASs for muscle traits, many studies have focused on a "responder" analysis of professional athletes (or Olympians) vs. control populations. Of course, nongenetic traits such as "effort and drive" have a major factor in becoming a professional athlete. That said, studies have shown some clear success stories. First and foremost, a nonsense polymorphism (an SNP that kills the gene and cannot make its corresponding protein) has been found in a fast-twitch myofiber isoform of alpha-actinin (a structural protein of the Z-line of the sarcomere that binds actin filaments) [14]. While the ACTN3 polymorphism (the gene for speed) was originally identified via candidate gene studies, its large effect size has held up in subsequent GWAS analyses [15]. The loss-of-function polymorphism is associated with endurance phenotypes, whereas the wild-type allele is associated with strength and power phenotypes.

Besides muscle function studied through phenotypes such as speed or strength, the amount of muscle mass may be studied by measuring lean mass (mostly consisting of skeletal muscle) by X-ray absorptiometry of electrical impedance. Collaborative GWASs of lean body mass in a total of 38,292 participants have identified five validated loci: *HSD17B11*, *VCAN*, *ADAMTSL3*, *IRS1*, and *FTO* [16]. As lean and fat mass are usually directly correlated, investigators adjusted lean mass association signal for fat mass, confirming the strong association of the *FTO* gene with both lean and fat mass regulation. Another area where GWASs have shed light on muscle physiology is type 2 diabetes, where it is well known that exercise is the best treatment for reversing this disorder [17].

2.3.2 Applications in Muscle Disease

Skeletal muscle diseases are challenging to investigate by GWAS because of their rarity, which hinders the collection of samples large enough to yield the required statistical power for GWAS. Even if one considered "muscular dystrophy" or "myopathy" as a single, large nosographic category, comprising all described forms together, GWAS would hardly be an effective method of exploring the underlying genetic variants. This is partly because of extreme locus heterogeneity, i.e., a very large number of genes responsible for the phenotype in different individuals, which would impede the identification of strong association signals, but mostly because the rare causative mutations of Mendelian myopathies are not efficiently tagged by common SNPs included in genotyping chips and may only be identified by sequencing. A GWAS approach, however, may be used to investigate the modifier effect of common genetic variation on phenotype in groups of individuals with the same neuromuscular disease, caused by identical or equivalent mutations within the same gene, leading to the identification of genetic modifiers. Acquired diseases such as inflammatory or toxic myopathies may also be investigated by GWAS, looking for common risk variants.

GWAS in Acquired Muscle Diseases The most important chapter of acquired muscle diseases is probably represented by the idiopathic inflammatory myopathies, whose pathogenesis bears resemblance to that of other autoimmune/ autoinflammatory diseases. In the GWAS era, much has been learned about the genetic architecture of autoimmune disorders. The human leukocyte antigen (HLA) complex, encoded in a 3-Mb region on chromosome 6p21, has long been known (well before the emergence of GWAS) to be associated with the risk of developing several autoimmune disorders. A challenge for autoimmune disease GWASs has been the fine-mapping of this region [18], which is very dense in implicated variants and association signals, leading to the identification of responsible amino acid variations through a combined use of haplotype-based imputation, targeted resequencing, and development of dedicated genotyping tools such as the Immunochip. Outside this genomic region, GWASs have brought to the identification of numerous other association signals near genes involved in inflammation and immunity [19]. A typical feature of the genetic architecture of these disorders is pervasive pleiotropy, many individual loci being associated with different autoimmune conditions and often acting paradoxically as risk or protective variants in different diseases, depending on the pathological context [20].

Inflammatory myopathies comprise the largest group of acquired and treatable muscle disorders in the pediatric and adult population and are divided in a few major forms: dermatomyositis (DM), polymyositis (PM), sporadic inclusion body myositis (sIBM), necrotizing autoimmune myopathy (NAM), and the "overlap" myositides which associate with other connectivitis conditions [21]. A strong association with the HLA 8.1 ancestral haplotype has been established in a GWAS including a mixed PM and DM population [22]. Differences between DM and PM have been identified in respect to both the fine-mapping of the association in the HLA region and independent association signals in other loci, as shown by Immunochip studies [23]. The *PTPN22* gene, encoding for a protein tyrosine phosphatase expressed in lymphoid tissues, and implicated in T-cell receptor signaling, seems to play a relevant role in PM, but not DM, pathogenesis. Validation studies across ethnic groups seem to confirm the strengths of these associations [24]. sIBM recognizes a different pathological picture, with a relevant degenerative component and an accordingly different genetic component. So far, only the HLA-DRB1*03:01 locus has been associated with sIBM at genome-wide level, and non-HLA associations have only been proposed as putative [25].

Another muscle-related condition that is a frequent reason of referral to neuromuscular centers is muscle toxicity due to the use of the commonly prescribed lipid-lowering agents, statins, which can take the form of isolated elevation of circulating creatine kinase (CK), myalgia, overt myopathy with muscle weakness, and severe rhabdomyolisis in rare cases [26]. An early GWAS enrolling only 85 subjects with statin-induced myopathy and 90 controls identified a strong association (OR 4.5) with the rs4149056 SNP within *SLCO1B1*, involved in the hepatic uptake of statins [27], while subsequent larger GWASs exploring CK values as a QT identified associated variants within the CK-encoding *CKM* gene, as well as several genes related with immune functions [28, 29].

Genetic Modifiers of Muscle Diseases Genetic modifiers are variants, situated in genes different from the disease gene of a Mendelian disorder, which by a *trans*-active mechanism influence its expressivity. Phenotypic variability is a common observation in myopathies and muscular dystrophies [30], representing a challenge for definition of the natural history and for clinical trial design [31]. The identification of genetic modifiers allows more precise prognosis and patient stratification in observational and interventional studies. Duchenne muscular dystrophy (DMD), the severe form of dystrophinopathy caused by frameshifting or truncating mutations in the *DMD* gene [32, 33], has represented a paradigm in the study of genetic modifiers in the muscle disease field, for several reasons: there is little phenotype variability due to different disease-causing mutations ("*cis*-effects," as opposed to genetic modifier "*trans*-effects") because by definition all DMD patients have mutations leading to a complete dystrophin defect; DMD is relatively frequent among the rare muscular dystrophies, facilitating the collection of larger case series; DMD is severe, allowing to observe and measure the evolution of muscle weakness more easily in the time frame of a clinical study. Several genetic modifier loci of DMD have been identified by a candidate gene approach. Osteopontin, encoded by the gene *SPP1*, is a pleiotropic cytokine known to be upregulated in damaged and dystrophic muscle, where it is expressed by both infiltrating immune cells and regenerating myoblasts [34]. Its genetic ablation in mice-delayed inflammation and repair after muscle damage [35] and apparently attenuated the damaging consequences of dystrophin deficiency [36, 37]. The minor G allele at the *SPP1* promoter SNP rs28357094, previously tied to a reduction in gene expression in in vitro models [38], was shown to be associated with earlier loss of ambulation (LoA) in collectively around 370 participants from the CINRG-DNHS and the DMD cohort followed at University of Padova, Italy [9, 39]. Participants in the CINRG-DNHS also showed lower grip strength with the G genotype, while participants from the Padova cohort had lower manual muscle testing strength scores. The association was validated with longitudinal ambulatory outcomes (North Star Ambulatory Assessment and 6-min walk test) in an independent Italian cohort [40]. On the other hand, validation was problematic in further DMD cohorts [41, 42], possibly due to a relatively small effect size and complex interactions of the *SPP1* promoter with steroids, including glucocorticoids used to treat DMD [9, 43, 44].

Latent transforming growth factor β (TGF-β) binding protein 4 (LTBP4) is a matrix protein that binds TGF-β in a latent complex in the extracellular matrix, preventing it from interacting with its receptor at the cellular surface. The murine *Ltbp4* locus emerged as a potential modifier from a preclinical genome scan of a mouse model of muscular dystrophy lacking δ-sarcoglycan, on a diverse genetic background [45]. Fine-mapping of this association revealed a 36-bp deletion in the *Ltbp4* coding region, associated with increased proteolysis of the TGF-β latent complex, downstream SMAD signaling, and fibrosis. A subsequent candidate-gene study in a cohort of 254 severe dystrophinopathy patients followed by the United Dystrophinopathy Project (UDP) revealed that a haplotype of four coding SNPs across different exons of the human homologous *LTBP4* gene was associated with DMD phenotype severity, the minor haplotype IAAM leading to increased stability of the latent complex, reduced TGF-β signaling, and delayed LoA [41]. This association was validated in several hundred patients across other independent cohorts: CINRG-DNHS, Padova University [9], and the European Bio-NMD network [42].

Interestingly, both these modifiers consistently pointed to a relevant role of TGF-β signaling in modulating the severity of dystrophic pathology, as had been suggested by previous independent gene expression studies [46, 47].

A third DMD modifier identified with a candidate approach was the common null polymorphism R577X in the *ACTN3* gene, encoding the sarcomeric protein actinin-3. This allele is detrimental to sprint and power performance in athletes and normal individuals and was shown to also reduce muscle strength and 10-m walk/run speed in DMD [48].

The application of GWAS to the identification of genetic modifiers of muscular dystrophy has only seen initial attempts so far. One hundred and seventy DMD patients participating in the CINRG-DNHS were genotyped with the Illumina Exome chip, a genotyping chip focused on coding variants situated within protein-coding genes across the genome and mostly with protein-altering effects (missense, nonsense, or variants in the promoters and untranslated regions [UTRs]). These exon variants were associated with the LoA phenotype (Cox regression with glucocorticoid treatment as a covariate) in 109 participants of European ancestry but did not yield "exome-wide" significant signals after Bonferroni correction for about 27,000 examined SNPs. A hypothesis-based filtering and prioritization of suggestive association signals, selecting those SNPs in the vicinities of genes involved in pro-inflammatory and pro-fibrotic pathways, singled out a signal corresponding to the rs1883832 SNP in the 5′ UTR of the *CD40* gene, encoding for a co-stimulatory T-cell receptor. The minor T allele at rs188382 had already been associated with different immune diseases both as a risk factor (for multiple sclerosis and Graves' disease) and conversely as a protective factor (e.g., for some vasculitides such as Kawasaki disease), as is often observed in pleiotropic loci associated with autoinflammation. Validation in around 550 DMD patients from all international DMD natural history studies cited above confirmed the association of the rs1883832 minor T allele with earlier LoA in DMD, highlighting the relevance of the transition from innate to cell-mediated immunity, mediated by T-cell receptor signaling, in DMD [10].

The first GWAS utilizing a "classic" genotyping chip (as opposed to the exome chip which focuses on exonic, disrupting variants) has recently been published by the UDP group [49]. The authors focused again on the LoA phenotype in the same cohort of 253 severe dystrophinopathy patients used to describe the LTBP4 modifier variant described above and, despite the small sample size for GWAS standards, were able to identify two different loci with genome-wide significance. The first significant SNP, rs2725797 (chr15, $p = 6.6 \times 10^{-9}$), was demonstrated through analyses of gene expression and chromatin interaction databases to tag regulatory variants relevant to the expression of thrombospondin-1. This protein, encoded by the *THBS1* gene, is an activator of TGF-β signaling by direct binding to LTBP4 and an inhibitor of pro-angiogenic nitric oxide signaling. The second signal derived from the SNP, rs710160 (chr19, $p = 4.7 \times 10^{-8}$), is situated upstream of the *LTBP4* gene and functionally related to its expression levels. This work provides further confirmation of the key role of the LTBP4-TGF-β pathway in dictating DMD pathology. Apparently, multiple SNPs situated in different chromosomes converge on this same molecular pathway. This work is also interesting for the analytic strategy used to compensate for small sample size: inclusion of familial cases with ad hoc computational algorithms to adjust for relatedness, implementation of different inheritance models (similar to the original LTBP4 association, both loci were significant with the recessive model), and in silico functional testing leveraging on open online databases of functional data on gene expression and chromatin conformation.

Especially the last two works here reviewed provide a proof of principle that, at least in the most common of rare diseases, it is possible to apply a GWAS approach to the study of variable phenotypic expressivity.

2.4 Limitations and Extensions

GWAS is a powerful tool that has provided much knowledge about the genetic bases of human diseases and traits. However, it has been clear for at least 10 years that a "classic" GWAS approach—genotyping hundreds/thousands of individuals for SNPs tagging common variation in the genome—is sufficient to explain but a small percentage of the heritability of these diseases and traits [7]. The "missing heritability" may be explained by several reasons: inefficient tagging of functional variants by genotyped SNPs; presence of large numbers of associated SNPs with very small effects that cannot be demonstrated, if not by huge meta-analyses of hundreds of thousands of individuals; effects of rare variants and copy number variations; and epigenetic changes. The continuous improvement of genotyping chips with dense genotyping and ethnicity-specific arrays will improve GWAS power, by more efficient tagging of common variation. Imputation of infrequent and rare variants can be effectively performed from haplotypes established in sequencing studies in homogeneous populations [50]. Low-frequency exon variants identified by WES and WGS studies in more than one population (exome chips) can be included in genotyping arrays. Analyses of signal intensity from genotyping assays can be performed to identify copy number variations.

Another great challenge is the investigation of the interaction of genetic variants with the environment and between themselves. The characterization of environmental interactions requires deep phenotyping and complex, versatile statistical models that are hard to fit to a genome-wide discovery study. Regarding epistatic gene x gene interactions, GWAS has been largely unsuccessful in demonstrating them. A possible reason is that in classic association tests, statistical power is proportional to the LD between the tag SNP and the functional variant, measured as r^2, but when two functional variants are involved, statistical power decreases proportionally to r^4 when tag SNPs do not perfectly predict the functional variant [8]. Sequencing data and imputation of rare variants, applied to large GWAS meta-analyses, may improve the power of studies of epistatic interactions in the future.

The field of rare variant association is rapidly evolving and has the potential to explain some of the "missing heritability" remaining from GWASs. This approach is very interesting for Mendelian diseases such as myopathies and muscular dystrophy, whose genetic architecture is largely weighted on rare, gene-disrupting variants. In fact, "gene hunting" in unsolved neuromuscular disorders by whole exome sequencing (WES) and, increasingly, by whole genome sequencing (WGS) may be considered a form of rare variant association, even if variants are tested for pathogenicity case by case, rather than collectively. Statistical tests applied to the effect of rare (MAF < 0.01 down to singleton) variants cannot obviously be based on a single-variant approach, but rather rely on "collapsing" identified variants within predefined genetic sequence unit, most conveniently (but not necessarily) protein-coding genes. The first such algorithms, called "burden" tests, leaned heavily on the hypothesis of disruptive disease-causing variants, with substantial loss of power in the presence of neutral variants and bidirectional effects [51]. Subsequent advancements provided researchers with algorithms that can be adapted to different scenarios, depending on various scientific hypotheses, such as those based on the sequence kernel association test (SKAT), and may be optimized for relatively small sample sizes [52].

2.5 Future Directions

Applications of GWAS to muscle disorders have yet been only initial. A large, adequately powered GWAS based on the LoA phenotype in DMD would require a sample size of around 800–1000 individuals according to our power calculations (Fig. 2.6) and is feasible in a multicenter setting. Identified loci may be cross-

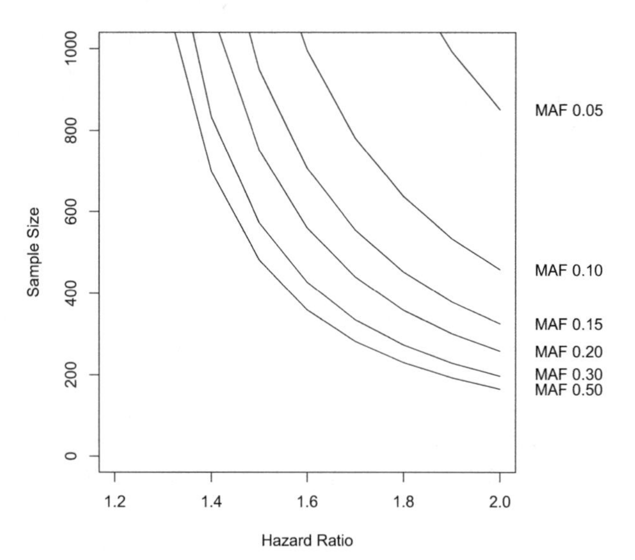

Fig. 2.6 Power calculation for a GWAS of age at LoA in DMD, assuming a type I error rate of 5×10^{-8} and power 0.8. The identification of moderate effects of common variants (effect sizes similar to those described for known genetic modifiers identified in candidate studies) require sample sizes around 800–1000 individuals

validated in other dystrophinopathies and muscular dystrophies in general, as several pathogenetic mechanisms might cross disease boundaries. Studying genetic modifiers in even rarer muscle disorders is challenging both because of sample sizes and because of the difficulties in modeling phenotype differences due to allelic heterogeneity (i.e., different pathogenetic mutations within the same gene causing the same disease).

The future of neuromuscular diagnostics may entail the integration of WES, or even WGS, and SNP genotyping, aiming primarily at the identification of the Mendelian pathogenetic mutation, but also simultaneously at defining and discovering modifier variants, paving the way for a personalized genomic medicine of muscle disorders.

References

1. Lander, E. S., Linton, L. M., Birren, B., Nusbaum, C., Zody, M. C., Baldwin, J., Devon, K., Dewar, K., Doyle, M., FitzHugh, W., et al. (2001). Initial sequencing and analysis of the human genome. *Nature, 409*, 860–921.
2. Venter, J. C., Adams, M. D., Myers, E. W., Li, P. W., Mural, R. J., Sutton, G. G., Smith, H. O., Yandell, M., Evans, C. A., Holt, R. A., et al. (2001). The sequence of the human genome. *Science, 291*, 1304–1351.
3. Sachidanandam, R., Weissman, D., Schmidt, S. C., Kakol, J. M., Stein, L. D., Marth, G., Sherry, S., Mullikin, J. C., Mortimore, B. J., Willey, D. L., et al. (2001). A map of human genome sequence variation containing 1.42 million single nucleotide polymorphisms. *Nature, 409*, 928–933.
4. International HapMap Consortium. (2005). A haplotype map of the human genome. *Nature, 437*, 1299–1320.
5. McCarthy, M. I., Abecasis, G. R., Cardon, L. R., Goldstein, D. B., Little, J., Ioannidis, J. P. A., & Hirschhorn, J. N. (2008). Genome-wide association studies for complex traits: Consensus, uncertainty and challenges. *Nature Reviews Genetics, 9*, 356–369.
6. Klein, R. J. (2005). Complement factor H polymorphism in age-related macular degeneration. *Science, 308*, 385–389.
7. Maher, B. (2008). Personal genomes: The case of the missing heritability. *Nature, 456*, 18–21.
8. Visscher, P. M., Wray, N. R., Zhang, Q., Sklar, P., McCarthy, M. I., Brown, M. A., & Yang, J. (2017). 10 years of GWAS discovery: Biology, function, and translation. *American Journal of Human Genetics, 101*, 5–22.
9. Bello, L., Kesari, A., Gordish-Dressman, H., Cnaan, A., Morgenroth, L. P., Punetha, J., Duong, T., Henricson, E. K., Pegoraro, E., McDonald, C. M., et al. (2015). Genetic modifiers of ambulation in the cooperative international neuromuscular research group Duchenne natural history study. *Annals of Neurology, 77*, 684–696.
10. Bello, L., Flanigan, K. M., Weiss, R. B., United Dystrophinopathy Project, Spitali, P., Aartsma-Rus, A., Muntoni, F., Zaharieva, I., Ferlini, A., Mercuri, E., et al. (2016). Association study of exon variants in the NF-κB and TGFβ pathways identifies CD40 as a modifier of Duchenne muscular dystrophy. *American Journal of Human Genetics, 99*, 1163–1171.
11. Fontanesi, L., Schiavo, G., Galimberti, G., Bovo, S., Russo, V., Gallo, M., & Buttazzoni, L. (2017). A genome-wide association study for a proxy of intermuscular fat level in the Italian Large White breed identifies genomic regions affecting an important quality parameter for dry-cured hams. *Animal Genetics, 48*, 459–465.

12. Jiang, Y., Tang, S., Wang, C., Wang, Y., Qin, Y., Wang, Y., Zhang, J., Song, H., Mi, S., Yu, F., Xiao, W., Zhang, Q., & Ding, X. (2018). A genome-wide association study of growth and fatness traits in two pig populations with different genetic backgrounds. *Journal of Animal Science, 96*, 806–816.
13. Yoshida, G. M., Lhorente, J. P., Carvalheiro, R., & Yáñez, J. M. (2017). Bayesian genome-wide association analysis for body weight in farmed Atlantic salmon (Salmo salar L.). *Animal Genetics, 48*, 698–703.
14. Eynon, N., Hanson, E. D., Lucia, A., Houweling, P. J., Garton, F., North, K. N., & Bishop, D. J. (2013). Genes for elite power and sprint performance: ACTN3 leads the way. *Sports Medicine, 43*, 803–817.
15. Ahmetov, I. I., Egorova, E. S., Gabdrakhmanova, L. J., & Fedotovskaya, O. N. (2016). Genes and athletic performance: An update. *Medicine and Sport Science, 61*, 41–54.
16. Zillikens, M. C., Demissie, S., Hsu, Y. H., Yerges-Armstrong, L. M., Chou, W. C., Stolk, L., Livshits, G., Broer, L., Johnson, T., Koller, D. L., et al. (2017). Large meta-analysis of genome-wide association studies identifies five loci for lean body mass. *Nature Communications, 8*, 80.
17. Gaulton, K. J. (2017). Mechanisms of type 2 diabetes risk loci. *Current Diabetes Reports, 17*, 72.
18. Spain, S. L., & Barrett, J. C. (2015). Strategies for fine-mapping complex traits. *Human Molecular Genetics, 24*, R111–R119.
19. Sorrentino, R. (2014). Genetics of autoimmunity: An update. *Immunology Letters, 158*, 116–119.
20. Parkes, M., Cortes, A., van Heel, D. A., & Brown, M. A. (2013). Genetic insights into common pathways and complex relationships among immune-mediated diseases. *Nature Reviews Genetics, 14*, 661–673.
21. Dalakas, M. C. (2015). Inflammatory muscle diseases. *The New England Journal of Medicine, 372*, 1734–1747.
22. Miller, F. W., Chen, W., O'Hanlon, T. P., Cooper, R. G., Vencovsky, J., Rider, L. G., Danko, K., Wedderburn, L. R., Lundberg, I. E., Pachman, L. M., et al. (2015). Genome-wide association study identifies HLA 8.1 ancestral haplotype alleles as major genetic risk factors for myositis phenotypes. *Genes and Immunity, 16*, 470–480.
23. Rothwell, S., Cooper, R. G., Lundberg, I. E., Miller, F. W., Gregersen, P. K., Bowes, J., Vencovsky, J., Danko, K., Limaye, V., Selva-O'Callaghan, A., et al. (2016a). Dense genotyping of immune-related loci in idiopathic inflammatory myopathies confirms HLA alleles as the strongest genetic risk factor and suggests different genetic background for major clinical subgroups. *Annals of the Rheumatic Diseases, 75*, 1558–1566.
24. Rothwell, S., Lamb, J. A., & Chinoy, H. (2016b). New developments in genetics of myositis. *Current Opinion in Rheumatology, 28*, 651–656.
25. Britson, K. A., Yang, S. Y., & Lloyd, T. E. (2018). New developments in the genetics of inclusion body myositis. *Current Rheumatology Reports, 20*(5), 26.
26. Hilton-Jones, D. (2018). Statin-related myopathies. *Practical Neurology, 18*, 97–105.
27. SEARCH Collaborative Group, Link, E., Parish, S., Armitage, J., Bowman, L., Heath, S., Matsuda, F., Gut, I., Lathrop, M., & Collins, R. (2008). SLCO1B1 variants and statin-induced myopathy—A genomewide study. *The New England Journal of Medicine, 359*, 789–799.
28. Dube, M.-P., Zetler, R., Barhdadi, A., Brown, A. M. K., Mongrain, I., Normand, V., Laplante, N., Asselin, G., Zada, Y. F., Provost, S., et al. (2014). CKM and LILRB5 are associated with serum levels of Creatine kinase. *Circulation. Cardiovascular Genetics, 7*, 880–886.
29. Kristjansson, R. P., Oddsson, A., Helgason, H., Sveinbjornsson, G., Arnadottir, G. A., Jensson, B. O., Jonasdottir, A., Jonasdottir, A., Bragi Walters, G., Sulem, G., et al. (2016). Common and rare variants associating with serum levels of creatine kinase and lactate dehydrogenase. *Nature Communications, 7*, 10572.
30. Humbertclaude, V., Hamroun, D., Bezzou, K., Bérard, C., Boespflug-Tanguy, O., Bommelaer, C., Campana-Salort, E., Cances, C., Chabrol, B., Commare, M.-C., et al. (2012). Motor and respiratory heterogeneity in Duchenne patients: Implication for clinical trials. *European Journal*

of Paediatric Neurology: EJPN: Official Journal of European Paediatric Neurology Society, 16, 149–160.
31. Ricotti, V., Muntoni, F., & Voit, T. (2015). Challenges of clinical trial design for DMD. *Neuromuscular Disorders, 25*, 932–935.
32. Hoffman, E. P., Brown, R. H., & Kunkel, L. M. (1987). Dystrophin: The protein product of the Duchenne muscular dystrophy locus. *Cell, 51*, 919–928.
33. Monaco, A. P., Bertelson, C. J., Liechti-Gallati, S., Moser, H., & Kunkel, L. M. (1988). An explanation for the phenotypic differences between patients bearing partial deletions of the DMD locus. *Genomics, 2*, 90–95.
34. Uaesoontrachoon, K., Yoo, H.-J., Tudor, E. M., Pike, R. N., Mackie, E. J., & Pagel, C. N. (2008). Osteopontin and skeletal muscle myoblasts: Association with muscle regeneration and regulation of myoblast function in vitro. *The International Journal of Biochemistry and Cell Biology, 40*, 2303–2314.
35. Uaesoontrachoon, K., Wasgewatte Wijesinghe, D. K., Mackie, E. J., & Pagel, C. N. (2013). Osteopontin deficiency delays inflammatory infiltration and the onset of muscle regeneration in a mouse model of muscle injury. *Disease Models and Mechanisms, 6*, 197–205.
36. Capote, J., Kramerova, I., Martinez, L., Vetrone, S., Barton, E. R., Sweeney, H. L., Miceli, M. C., & Spencer, M. J. (2016). Osteopontin ablation ameliorates muscular dystrophy by shifting macrophages to a pro-regenerative phenotype. *The Journal of Cell Biology, 213*, 275–288.
37. Vetrone, S. A., Montecino-Rodriguez, E., Kudryashova, E., Kramerova, I., Hoffman, E. P., Liu, S. D., Miceli, M. C., & Spencer, M. J. (2009). Osteopontin promotes fibrosis in dystrophic mouse muscle by modulating immune cell subsets and intramuscular TGF-β. *The Journal of Clinical Investigation, 119*, 1583–1594.
38. Giacopelli, F., Marciano, R., Pistorio, A., Catarsi, P., Canini, S., Karsenty, G., & Ravazzolo, R. (2004). Polymorphisms in the osteopontin promoter affect its transcriptional activity. *Physiological Genomics, 20*, 87–96.
39. Pegoraro, E., Hoffman, E., Piva, L., Gavassini, B., Cagnin, S., Ermani, M., Bello, L., Soraru, G., Pacchioni, B., Bonifati, M., et al. (2011). SPP1 genotype is a determinant of disease severity in Duchenne muscular dystrophy. *Neurology, 76*, 219–226.
40. Bello, L., Piva, L., Barp, A., Taglia, A., Picillo, E., Vasco, G., Pane, M., Previtali, S. C., Torrente, Y., Gazzerro, E., et al. (2012). Importance of SPP1 genotype as a covariate in clinical trials in Duchenne muscular dystrophy. *Neurology, 79*, 159–162.
41. Flanigan, K. M., Ceco, E., Lamar, K.-M., Kaminoh, Y., Dunn, D. M., Mendell, J. R., King, W. M., Pestronk, A., Florence, J. M., Mathews, K. D., et al. (2013). LTBP4 genotype predicts age of ambulatory loss in Duchenne muscular dystrophy. *Annals of Neurology, 73*, 481–488.
42. van den Bergen, J. C., Hiller, M., Böhringer, S., Vijfhuizen, L., Ginjaar, H. B., Chaouch, A., Bushby, K., Straub, V., Scoto, M., Cirak, S., et al. (2015). Validation of genetic modifiers for Duchenne muscular dystrophy: A multicentre study assessing SPP1 and LTBP4 variants. *Journal of Neurology, Neurosurgery, and Psychiatry, 86*, 1060–1065.
43. Hoffman, E. P., Gordish-Dressman, H., McLane, V. D., Devaney, J. M., Thompson, P. D., Visich, P., Gordon, P. M., Pescatello, L. S., Zoeller, R. F., Moyna, N. M., et al. (2013). Alterations in osteopontin modify muscle size in females in both humans and mice. *Medicine and Science in Sports and Exercise, 45*, 1060–1068.
44. Vianello, S., Pantic, B., Fusto, A., Bello, L., Galletta, E., Borgia, D., Gavassini, B. F., Semplicini, C., Sorarù, G., Vitiello, L., et al. (2017). SPP1 genotype and glucocorticoid treatment modify osteopontin expression in Duchenne muscular dystrophy cells. *Human Molecular Genetics, 26*, 3342–3351.
45. Heydemann, A., Ceco, E., Lim, J. E., Hadhazy, M., Ryder, P., Moran, J. L., Beier, D. R., Palmer, A. A., & McNally, E. M. (2009). Latent TGF-beta-binding protein 4 modifies muscular dystrophy in mice. *The Journal of Clinical Investigation, 119*, 3703–3712.

46. Chen, Y.-W., Nagaraju, K., Bakay, M., McIntyre, O., Rawat, R., Shi, R., & Hoffman, E. P. (2005). Early onset of inflammation and later involvement of TGF in Duchenne muscular dystrophy. *Neurology, 65*, 826–834.
47. Dadgar, S., Wang, Z., Johnston, H., Kesari, A., Nagaraju, K., Chen, Y.-W., Hill, D. A., Partridge, T. A., Giri, M., Freishtat, R. J., et al. (2014). Asynchronous remodeling is a driver of failed regeneration in Duchenne muscular dystrophy. *The Journal of Cell Biology, 207*, 139–158.
48. Hogarth, M. W., Houweling, P. J., Thomas, K. C., Gordish-Dressman, H., Bello, L., Cooperative International Neuromuscular Research Group (CINRG), Pegoraro, E., Hoffman, E. P., Head, S. I., & North, K. N. (2017). Evidence for ACTN3 as a genetic modifier of Duchenne muscular dystrophy. *Nature Communications, 8*, 14143.
49. Weiss, R. B., Vieland, V. J., Dunn, D. M., Kaminoh, Y., Flanigan, K. M., & United Dystrophinopathy Project. (2018). Long-range genomic regulators of THBS1 and LTBP4 modify disease severity in duchenne muscular dystrophy. *Annals of Neurology, 84*, 234–245.
50. Hoffmann, T. J., & Witte, J. S. (2015). Strategies for imputing and analyzing rare variants in association studies. *Trends in Genetics, 31*, 556–563.
51. Morris, A. P., & Zeggini, E. (2010). An evaluation of statistical approaches to rare variant analysis in genetic association studies. *Genetic Epidemiology, 34*, 188–193.
52. Lee, S., Abecasis, G. R., Boehnke, M., & Lin, X. (2014). Rare-variant association analysis: Study designs and statistical tests. *American Journal of Human Genetics, 95*, 5–23.

Chapter 3
Chromatin Epigenomics in Muscle Development and Disease

Jelena Perovanovic

3.1 Introduction

Eukaryotic genetic material is separated from the rest of the cytoplasm by a nuclear envelope. This unique feature distinguishes eukaryotes from prokaryotes and plays a role in cell specification and differentiation. Within the nucleus, genomes are hierarchically packed into the chromatin [1] that is tightly organized to provide a platform for regulation of gene expression. Histones are basic proteins, and their positive charges associate with negatively charged DNA to form chromatin. Chromatin is organized as an array of nucleosomes that are composed of histone octamer core (H2A, H2B, H3, H4). The first level of DNA packaging represents wrapping of 147 bp of DNA around this core. Each core is separated by linker DNA (H1) [2, 3]. By modulating the levels of this interaction, eukaryotic genomes regulate gene expression.

Chromatin is further organized into higher-order structures that not only allow packaging into the nucleus but also provide a regulatory mechanism [4]. Regulation of chromatin packaging occurs at various levels. Nucleosome localization has a negative effect on the transcription of the genes in the vicinity. Nucleosomes present a physical obstruction that can lead to bending of the DNA that impedes transcription. Furthermore, positioning of the nucleosomes along the DNA fiber has a major role in modifying accessibility of the specific biding sites to the corresponding transcription factors [5]. The tail region of core histone proteins can be covalently modified (Fig. 3.1). There are numerous residues within that tail that can be posttranslationally modified. New covalent modifications of the histone tails are still being discovered; however, the best studied ones occur at the N-terminus and include methylation (Me), acetylation (Ac), phosphorylation (P), ubiquitinylation, sumoylation, ADP ribosylation, and deamination [6]. Histone methylations occur on

J. Perovanovic (✉)
Huntsman Cancer Institute, University of Utah School of Medicine, Salt Lake City, UT, USA
e-mail: jelena.perovanovic@hci.utah.edu

J. G. Burniston, Y.-W. Chen (eds.), *Omics Approaches to Understanding Muscle Biology*, Methods in Physiology, https://doi.org/10.1007/978-1-4939-9802-9_3

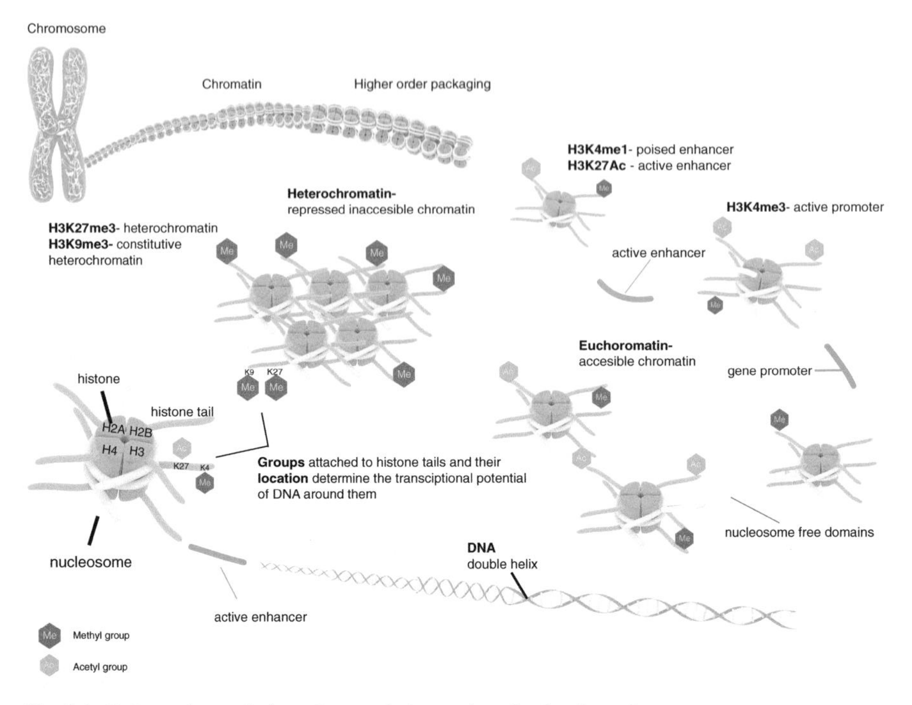

Fig. 3.1 Epigenetic regulation of transcription and packaging into chromosomes

lysines (K) and arginines (R), and acetylation takes place on lysines (K), while phosphorylation marks serine (S), threonine (T), and tyrosine (Y) residues [7]. Histone modifications can have both positive and negative effects on the transcription of the neighboring genes (Fig. 3.1). The tail residue that is modified together with the type of covalent modification is important in determining the effects of modification and is referred to as histone code. Not all histone modifications are important for regulation of gene expression; some modifications are constitutive and are an essential part of DNA repair, replication, and cell fate maintenance [8]. For example, histone phosphorylation plays a major role during DNA damage, when phosphorylated histone H2A(X) marks large chromatin domains around the site of DNA breakage [9].

Here we review major epigenomic techniques to study chromatin biology in muscle tissue with a major focus on muscle development, muscle regeneration, and aging.

3.2 Chromatin Immunoprecipitation

Chromatin immunoprecipitation (ChIP) is a commonly utilized method of defining protein/DNA interactions. Chromatin immunoprecipitation is performed on formaldehyde-fixed (cross-linked) cells or tissue. During this process a covalent bond is introduced between the protein of interest and the genomic regions bound by the protein. Using an antibody that specifically recognizes the protein of interest, the genomic loci directly or indirectly bound by the protein are isolated in the process called immunoprecipitation.

ChIP-seq combines ChIP with next-generation sequencing [10–12] to generate genome-wide maps of specific genomic regions that interact with a particular protein of interest. Next-generation sequencing is a high-throughput sequencing method that encompasses whole genome and exome sequencing, transcriptome profiling (RNA-seq), and epigenomic sequencing (ChIP-seq, ATAC-seq, etc.). The cost of high-throughput sequencing has dramatically decreased over the past several years making this technique more widely accessible.

Histone modifications have a pivotal role in influencing gene expression. Specific regulatory regions show unique epigenomic signatures. For example, trimethylation of lysine 4 on histone 3 (H3K4me3) is commonly found near transcription start sites, and it is a marker the promoters of transcriptionally active genes. Other posttranslational modifications are markers that are specific for enhancer regions (H3K27Ac, H3K4me1) or show negative correlation with transcriptional activity (H3K27me3, H3K9me3). These epigenetic signatures show high degrees of tissue specificity, and efforts have been made to generate comprehensive epigenome maps in multiple human and mouse tissues [Roadmap Epigenomics (http://www.roadmapepigenomics.org/)].

ChIP-seq is a powerful and versatile tool and there are many applications of this tool. Some of the major applications of ChIP-seq include analysis of chromatin modifications [13–15], chromatin state maps that define chromatin landscapes of normal and disease cells [16], analysis of transcription factor binding site [17, 18], and associations of architectural proteins with the DNA [19].

Bivalent domains are regulatory regions that are marked by large regions of H3 lysine 27 trimethylation (H3K27e3) and smaller regions of H3 lysine 4 trimethylation (H3K4me3) and were discovered using ChIP-seq [13]. In embryonic stem (ES) cells, bivalent domains mark developmental genes expressed at very low levels. These promoters tend to resolve during differentiation and either become enriched only for K27 methylation (lose K4 methylation) or keep only K4 methylation and drive silencing or activation of developmental genes, respectively [13]. With respect to the muscle field, Liu et al. investigated the bivalent epigenetic landscape of adult muscle progenitor cells known as quiescent satellite cells (QSCs) [15] and showed that chromatin of QSC is largely permissive and large numbers of promoters for non-myogenic lineage genes were marked as bivalent and very few were strictly repressed. During aging, QSCs acquire repressive K27 methylation marks while maintaining comparable levels of K4 methylation. These data provide evidence that epigenetic changes accumulate over the life span of an organism and represent a platform for functional decay of adult satellite cells in aging and/or pathological states [20, 21]. Brack et al. have shown that QSCs from aged mice tend to transdifferentiate from myogenic fate to fibrogenic lineage, and this is under the influence of the systemic environment of old animals [20]. This study highlights the importance of an aging environment and its contribution to pathological conditions (increase in tissue fibrosis with age).

Classical applications of ChIP-seq include identification of transcription factor binding sites, chromatin remodeling complex binding, and architectural protein-DNA interactions [18, 19, 22]. MyoD is a master regulator of muscle fate and plays a major role in the activation of skeletal muscle genes during differentiation. Using ChIP-seq, Cao et al. showed that MYOD1 binds many of the regulatory regions of muscle-specific genes and that this enrichment was increased with muscle differentiation [17]. The study confirmed MYOD1 as a master regulator of muscle-specific genes. Master regulators mediate cell signaling and direct cell-type-specific responses to environmental cues. Mullen et al. showed that transforming growth factor beta (TGF-β) signaling conveys its cell-type-specific response through SMAD3 master regulator interactions. The authors show that, in myogenic cells, Smad3 co-occupies genomic regions already bound by MYOD, whereas in ESC and B-cell-type cells, SMAD3 was enriched in OCT4 and PU.1 regions, respectively [18]. Finally, ChIP-seq technology has been utilized to study DNA interactions with architectural proteins. Lamin A/C protein is a major building block of nuclear lamina structure that lies underneath the nuclear membrane and is in direct contact with chromatin. In human adipose tissue, lamin A/C interacts with distinct promoter regions and associates with a repressive chromatin environment [19].

Chromatin profiling is an extension of classical ChIP-seq analysis of histone markers and provides a systematic description of *cis*-regulatory elements known as a chromatin state map. *Cis*-regulatory elements are regulatory regions that are present on the same molecule of DNA as the gene they control. Chromatin state maps use combinatorial analysis of various histone markers and DNA-binding proteins to define regulatory regions that are unique to particular cell types [16, 23]. Chromatin state maps define highly plastic landscapes, where specific patterns of remodeling are conserved among cell types. These maps use coloring scheme to show various genomic annotations in a clear and concise manner. For example, gray represents constitutive heterochromatin (Fig. 3.2), red and purple mark active and poised promoters, shades of yellow define enhancer regions, and green markers represent transcribed regions (Fig. 3.3) [24]. Moreover, chromatin states can be read in an allele-specific manner where disease variants frequently reside with *cis*-regulatory regions that are cell-type and tissue-type specific. In one such study [25], the authors analyzed muscle biopsies from 271 well-phenotyped Finnish participants ranging from normal to T2D. They showed that muscle-specific genetic regulatory architecture is highly enriched in muscle super enhancers. More specifically, they show that type 2 diabetes (T2D)-risk variants coincide with muscle super enhancers and are associated with increased expression of muscle-specific gene isoforms [25].

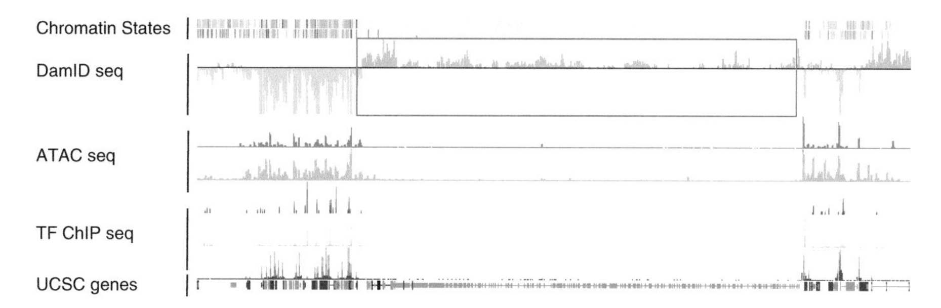

Fig. 3.2 Representative image of different sequencing resolutions generated using epigenetics tools. Top panel shows chromatin state maps in two cell types (brain, top; heart, bottom). Second panel shows the sequencing signal obtained though Dam-LMNB1 sequencing. The red box marks the DamID region known as LAD. Third panel represents ATAC-seq signal in two different cell lines. Fourth panel depicts signal generated by two classical transcription factors' binding (bright red and green) and histone modification pull-down (dark red). Bottom panel shows USCS-annotated gene positions

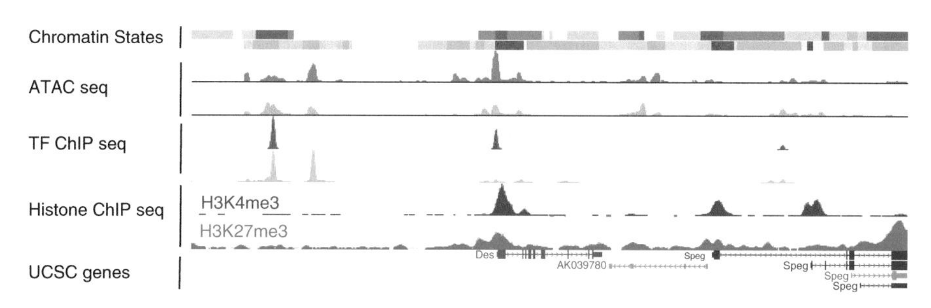

Fig. 3.3 Chromatin state maps, ATAC-seq signal, and ChIP-seq binding patterns are shown. Top panel shows chromatin state maps in two different tissues (brain, top; heart, bottom). Second panel depicts ATAC-seq signal from two different developmental stages. Third panel shows classical transcription factor binding patterns, followed by a fourth panel that represents enrichment of activating histone mark (H3K4me3 shown in dark red) and repressive histone modification (H3K27me3 shown in blue). Bottom panel shows locations of USCS-annotated genes

3.3 DamID

DNA adenine methyltransferase identification (DamID) assay is a technique that has been successfully used to study in vivo genome binding sites of a variety of DNA-binding and chromatin proteins [26–29]. DamID relies on prokaryotic adenine methyltransferase Dam that methylates adenines within GATC sequence that are specifically recognized by DpnI enzyme [29, 30]. When Dam is fused to a protein of interest, direct interaction of that protein and DNA brings the bacterial Dam enzyme in close proximity of the eukaryotic genome leading to methylation of the GATC sites nearby (a modification not normally seen in eukaryotic DNA). Thus, by identifying the newly methylated GATC sites, DamID provides a direct method for studying protein-DNA interactions.

The major difference between more commonly used ChIP and DamID is the method of sample preparation and the genomic regions of interest that can be identified. ChIP requires formaldehyde cross-linking of the sample to introduce covalent bonds between the target protein and bound DNA, thus identifying more stable interactions. DamID is an enzymatic modification of DNA that is more capable of detecting transient protein/DNA interactions as well as stable ones. The second major difference is in the way these two techniques identify genomic domains that are enriched for the protein of interest. ChIP requires the use of antibodies against the protein of interest, followed by isolation of the cross-linked DNA/protein complexes. The procedure is dependent on the quality of the antibody used and the ability to pull down low-solubility heterochromatic domains. The cross-linking procedure in ChIP can also induce sequencing artifacts [31].

DamID avoids these artifacts by direct methylation of the genomic DNA. Furthermore, since DamID is not dependent on an antibody, it is more broadly applicable to proteins where adequate antibodies may not exist [32]. DamID is also useful for testing the effects of mutations on DNA binding or for studying different isoforms of the same protein [33]. DamID has been successfully used to study nuclear envelope associations with the genome [27, 30, 33, 34]. These lamina-associated domains (LADs) have been first described by Guelen et al., as gene-poor regions, indicating that LADs are a repressive chromatin environment [33, 35]. Perovanovic et al. used the DamID system to analyze genome binding of different disease-causing mutant forms of lamin A/C protein [27, 36, 37]. The authors showed that muscle-specific mutation of lamin A causes inadequate chromatin tethering to the nuclear envelope leading to inappropriate activation of pluripotency pathway that resulted in delayed muscle differentiation [27]. Also, DamID has been used in the context of muscle to show that tissue-specific gene repositioning mediated by nuclear membrane promotes repression of developmental genes during myogenesis [34]. The authors provide evidence that three muscle-specific nuclear envelope transmembrane proteins, NET39, Tmem38A, and WFS1, direct specific myogenic genes to the nuclear periphery in order to facilitate their repression.

As shown in Fig. 3.2, DamID-positive signal spans over a large number of genes (represented by red box) and is positively correlated with heterochromatic regions (top chromatin state panel shows gray region that represents constitutive heterochromatin). Moreover, DamID-positive regions show closed chromatin architecture and are void of transcription binding sites and active histone marks.

3.4 ATAC-seq

Assay for transposase-accessible chromatin using sequencing (ATAC-seq) is a technique that provides an insight to the chromatin accessibility landscape. Nucleosome shifting or destabilization at the promoters or enhancer regions is positively correlated with transcription that results from binding of the specific transcription factors [38]. Thus the open and accessible genomic regions are considered to be primary locations of regulatory elements and have a determining role in cell fate specification, development, and disease progression. Further unraveling of the chromatin landscape remodeling is important to understand the mechanism of epigenomic control of normal development and disease progression. Representative ATAC-seq signals together with classical ChIP-seq enrichment from transcription factors or histone modifications are shown in Fig. 3.2.

ATAC-seq measures direct effects of chromatin remodeling on gene expression. It uses a process called tagmentation during which the genome is simultaneously fragmented and tagged by sequencing adaptors. Tagmentation is achieved by the use of hyperactive Tn5 transposase that can transpose DNA sequencing adapters into the native open chromatin. Mutant Tn5 is preloaded with the sequencing adapters, and the excision and DNA tagging occur in the same step. Initially, tagmentation was designed as part of commercial next-generation sequencing (NGS) library preparation kit. Since then this technology has been adapted to efficiently identify open chromatin regions, and it is now known as ATAC-seq [39].

ATAC-seq is a two-step protocol that uses between 500 and 50,000 cells to investigate chromatin accessibility, transcription factor occupancy, and nucleosome positioning. The major advantage of the ATAC-seq method over other methods that investigate the chromatin landscape (DNase-seq, FARE-seq) is the low amounts of starting material. DNase-seq relies on the DNase I enzyme to preferentially cleave open accessible domain sensitive to DNase I [38]. FARE-seq uses formaldehyde cross-linking to separate nucleosome bound (cross-linked) from nucleosome-free and accessible region (less cross-linked) [40]. Both methods require between 10^6 and 10^7 cells as a starting material. This is of particular interest to the muscle biology field where cell numbers can be a limiting factor. By requiring 50,000 freshly isolated cells instead of millions of cells grown ex vivo, ATAC-seq method preserves the in vivo epigenetic landscape and provides a tool to dissect individual populations within the muscle (i.e., quiescent satellite cells).

ATAC-seq tends to correlate with tissue-specific transcription factor binding and active H3K4me3 marks. In Fig. 3.2, desmin (Des) promoter shows an open chromatin profile (second panel) in two muscle cell types (purple, satellite cells; yellow, proliferating myoblasts). Furthermore, desmin promoter shows enrichment for muscle-specific MyoD (red) and active histone mark (H3K4me3) but not Pbx1 (green) and H3K27me3 (blue).

Major applications of ATAC-seq include nucleosomal mapping [39, 41], transcription factor occupancy [42, 43], identification of novel enhancers [44], and identification of accessibility changes relevant to pathological states [25].

Deep enough sequencing of ATAC-seq libraries coupled with computational analysis has a potential to infer the protein occupancy at specific nucleotides, known as footprinting [39]. The theory being that DNA sequences directly occupied by transcription factor (or other DNA-binding proteins) are inaccessible for transposition but are surrounded by accessible active chromatin. Computational identifications of these protected dips within the ATAC-seq regions could provide an alternative to numerous transcription factor ChIP-seq analyses in detailed reconstruction of regulatory networks. The idea of using digital footprinting is very appealing when considering limitations and biases introduced by ChIP-seq analysis (input requirements, cross-linking artifacts, antibody selection issues). Nevertheless, digital footprints have limitations. Not all transcription factors leave footprints [45], and it is highly dependent on individual transcription factor residence time [46].

ATAC-seq approach has been extensively used in the field of developmental biology to understand epigenetic remodeling. Wu et al. investigated chromatin remodeling in mammalian implantation embryos and showed that accessible chromatin is shaped by transposable elements and overlapped with *cis*-regulatory regions [43]. Using distal ATAC-seq peaks, the authors showed that in early-stage embryos, specific peaks were associated with genes controlling chromatin regulation, whereas distal peaks in inner cell mass (ICM) and embryonic stem cells (ESC) are located in the vicinity of development genes. They show that timing of the appearance of a particular motif correlates with the expression of the corresponding transcription factor. Furthermore, Loh et al. used ATAC-seq coupled with transcriptomic analysis to generate comprehensive developmental roadmaps of human development from pluripotency to 12 mesodermal lineages including skeletal and cardiac muscle. Using a stepwise process, they defined key transcriptional and signaling factors responsible for each lineage choice [42]. For example, the authors showed that activation of the dermomyotome program (embryonic stage that derives skeletal muscle) and inhibition of sclerotome (bone) depended on WNT activation and initial inhibition of hedgehog (HH) signaling. Chromatin accessibility maps identified GATA, HAND1, and NKX2.5 as key factors regulating cardiac mesoderm. These factors are considered as key biomarkers of cardiac tissue development. Finally, these data provide tools to navigate mesoderm and produce human tissue progenitors that could be used in therapy in addition to contributing to the fundamental understanding of human development.

ATAC-seq approach has been successfully applied to study disease states. Scott et al. analyzed skeletal muscle biopsies from 271 participants with glucose tolerance ranging from normal to newly diagnosed type 2 diabetes. Using ATAC-seq approach on frozen muscle sections, the authors investigated the tissue-specific genetic regulatory landscape and showed that type 2 diabetes risk variants were associated with muscle super enhancers and led to increased expression and alternative splicing of muscle-specific ANK1 isoforms [25].

3.5 Limitations

Next-generation sequencing technologies have been adapted to investigate different aspects of chromatin biology. These epigenomic tools have been successfully exploited to identify genomic loci that are bound by transcription factors, occupied by nucleosomes or accessible to nuclease cleavage or in close proximity to DamFusion constructs.

There are some key limitations that are associated with different steps of the process. Sample type and sample size represent a major limitation when it comes to high-throughput sequencing techniques. Muscle stem cells, quiescent satellite cells (QSCs), represent a small percentage of the muscle mass, and obtaining sufficient material for omics approach can be challenging. Traditional ChIP-seq protocols use between one million and ten million cells as starting material which is very challenging to collect when working with satellite cells where most isolation protocols yield between 100,000 and 350,000 quiescent SCs from hind limb muscles [47, 48]. To circumvent the sample size challenge, ChIP-seq protocols have been modified to work with low cell number. One such protocol, termed chipmentation, combines classical immunoprecipitation protocols with sequencing library preparation by Tn5 transposase to decrease input requirements to below 500,000 cells [49].

Chromatin structure represents the first level of bias in chromatin profiling studies. In ChIP-seq, DNA fragmentation is usually achieved by sonication and is a required step before protein-bound fragments are isolated by immunoprecipitation. The mechanical properties of the chromatin and its susceptibility to shearing vary across the genome, and it leads to a nonrandom selection of the fragments that are available for subsequent processing. Heterochromatin that contains tightly packed chromatin is less efficiently sheared compared to accessible euchromatin and usually underrepresented in sequencing studies. To circumvent this effect, enzymatic shearing is used in DamID studies and ATAC-seq studies, but this approach also has limitations and has been shown to be sequence dependent and represents a major source of bias.

The characteristics of transcription factor (TF) binding to the chromatin are substantially different between TFs [50]. The type of signals that are obtained from each TF is majorly influenced by nucleosomal positioning, kinetics of the TF-DNA interaction, and the TF tendency to co-bind with other factors [51]. Thus, some transcription factors are easily detected by ATAC-seq (e.g., CTCF), while other factors would never leave a footprint [46].

Resolution of each sequencing technique varies, and it should be taken into consideration in order to better understand the data and potential bias introduced by the experiment. Genomic analyses are performed over length scales from 1 bp (in SNP analyses) to around 10 bp (in ATAC-seq), 100 bp (in TF ChIP-seq), 100 bp–100 kb (in chromatin domain analyses—histone modification ChIP-seq), and 100 kb–1 Mb (DamID-seq) (Fig. 3.2).

The DamID method also has a unique set of limitations [36]. First, the fusion of the bacterial enzyme to the protein of interest may alter the activity and/or

DNA-binding properties of the resulting fusion protein. Second, DamID is not suitable for studying posttranslational modifications (PTMs) of the target protein. This is in contrast to ChIP that often uses PTM-specific antibodies to identify genomic regions of enrichment. DamID is an enzymatic modification of DNA that occurs over the entire time period of the experiment, making DamID less ideal for dissecting changes at a specific point in time. Additionally, prokaryotic Dam enzyme gets diluted every time eukaryotic cell replicates leading to loss of Dam methylation events over time. This limits the DamID applications in fast replicating cells (early embryonic development) [52]. Comparative studies have shown that both DamID and ChIP techniques produce comparable genome-wide binding profiles, suggesting that many of these potential limitations may not have a large impact on the results and data interpretation [33, 53].

3.6 Vital Future Directions

The vast majority of epigenetic data from cell cultures or tissues are generated and analyzed under the assumption that cells in culture or tissues are a homogeneous population that is often very far from reality. For example, epigenetic modifications are defined as transcriptionally activating or repressing based on experiments made in bulk cell populations. However, growing evidence contradicts this assumption and provides evidence for the greater complexity of epigenomic regulation [54]. Therefore, to better understand the complex relationship between epigenome and transcriptome in a heterogeneous setting, the epigenomic field is moving towards the single-cell approach.

The first single-cell (sc) ATAC-seq protocol was conducted by employing a "combinatorial indexing" approach in which the tagmentation reaction (and first barcoding) is performed in 96 wells containing a few thousand nuclei. During the second step, cells were pooled and split after which a second barcode was introduced by polymerase chain reaction (PCR). This experimental design was optimized in order to maximize the probability that a barcode combination marks only a single cell [55, 56]. Other scATAC-seq methods have been described that rely on commercially available microfluidic devices to carry out the transposition reaction in individual cells [57, 58].

Conducting ChIP-seq at the single-cell resolution is very challenging due to high levels of background noise associated with nonspecific antibody binding. Rotem et al. presented a modified protocol to overcome this challenge [59]. The authors performed the immunoprecipitation step on chromatin from pooled single cells that had already been fragmented (digested) and barcoded, so that the pull-down is effectively performed on thousands of cells. Large numbers of cells had to be processed in parallel using droplet-based microfluidic platform to obtain a sufficient number of high-quality reads per cell.

A single-cell approach has been successfully employed to study interactions with the nuclear lamina using single-cell DamID [60]. One of the limitations of the technique is the resolution which is in the order of 100 kb–2 Mb, which limits the extent of its applications.

The epigenomic field and tools are evolving at a remarkable pace due to advances in DNA sequencing technologies, accessibility of epigenomic data in the public domain [61], and the early adoption of new technologies into individual laboratories. These methods have challenges and limitations, such as high levels of PCR duplicates, sample size requirements, and introduction of technical bias (shearing bias, nonspecific antibody binding, lack of adequate controls). Nonetheless, single-cell sequencing tools provide new perspective into genomic variations that may underlie different disease states. Moreover, computational integration of these tools into multi-omics approach provides a platform for multiple data sets to be used for discovery and better understating of genome-cell function interplay. Understanding links between various epigenetic layers provides a window to comprehensively unlock the secrets of cell identity, function, and the progression of diseases [62].

References

1. Kornberg, R. D. (1974). Chromatin structure: A repeating unit of histones and DNA. *Science, 184*, 868–871.
2. Luger, K., Mäder, A. W., Richmond, R. K., Sargent, D. F., & Richmond, T. J. (1997). Crystal structure of the nucleosome core particle at 2.8 Å resolution. *Nature, 389*, 251–260.
3. Richmond, T. J., & Davey, C. A. (2003). The structure of DNA in the nucleosome core. *Nature, 423*, 145–150.
4. Woodcock, C. L., & Dimitrov, S. (2001). Higher-order structure of chromatin and chromosomes. *Current Opinion Genetics and Development, 11*, 130–135. http://www.ncbi.nlm.nih.gov/pubmed/11250134.
5. Radman-Livaja, M., & Rando, O. J. (2010). Nucleosome positioning: How is it established, and why does it matter? *Developmental Biology, 339*, 258–266.
6. Kouzarides, T. (2007). Chromatin modifications and their function. *Cell, 128*, 693–705.
7. Bannister, A. J., & Kouzarides, T. (2011). Regulation of chromatin by histone modifications. *Cell Research, 21*, 381–395.
8. Lawrence, M., Daujat, S., & Schneider, R. (2016). Lateral thinking: How histone modifications regulate gene expression. *Trends in Genetics, 32*(1), 42–56. https://doi.org/10.1016/j.tig.2015.10.007.
9. Rossetto, D., Avvakumov, N., & Côté, J. (2012). Histone phosphorylation: A chromatin modification involved in diverse nuclear events. *Epigenetics, 7*, 1098–1108.
10. Barski, A., Cuddapah, S., Cui, K., Roh, T.-Y., Schones, D. E., Wang, Z., Wei, G., Chepelev, I., & Zhao, K. (2007). High-resolution profiling of histone methylations in the human genome. *Cell, 129*, 823–837.
11. Johnson, D. S., Mortazavi, A., Myers, R. M., & Wold, B. (2007). Genome-wide mapping of in vivo protein-DNA interactions. *Science, 316*, 1497–1502.
12. Zhou, V. W., Goren, A., & Bernstein, B. E. (2011). Charting histone modifications and the functional organization of mammalian genomes. *Nature Reviews Genetics, 12*, 7–18.
13. Bernstein, B. E., Mikkelsen, T. S., Xie, X., Kamal, M., Huebert, D. J., Cuff, J., Fry, B., Meissner, A., Wernig, M., Plath, K., Jaenisch, R., Wagschal, A., Feil, R., Schreiber, S. L., & Lander, E. S. (2006). A bivalent chromatin structure marks key developmental genes in embryonic stem cells. *Cell, 125*, 315–326.

14. Hawkins, R. D., Hon, G. C., Lee, L. K., Ngo, Q., Lister, R., Pelizzola, M., Edsall, L. E., Kuan, S., Luu, Y., Klugman, S., Antosiewicz-Bourget, J., Ye, Z., Espinoza, C., Agarwahl, S., Shen, L., Ruotti, V., Wang, W., Stewart, R., Thomson, J. A., Ecker, J. R., & Ren, B. (2010). Distinct epigenomic landscapes of pluripotent and lineage-committed human cells. *Cell Stem Cell, 6*, 479–491.
15. Liu, L., Cheung, T. H., Charville, G. W., Hurgo, B. M. C., Leavitt, T., Shih, J., Brunet, A., & Rando, T. A. (2013). Chromatin modifications as determinants of muscle stem cell quiescence and chronological aging. *Cell Reports, 4*, 189–204.
16. Ernst, J., Kheradpour, P., Mikkelsen, T. S., Shoresh, N., Ward, L. D., Epstein, C. B., Zhang, X., Wang, L., Issner, R., Coyne, M., Ku, M., Durham, T., Kellis, M., & Bernstein, B. E. (2011). Mapping and analysis of chromatin state dynamics in nine human cell types. *Nature, 473*, 43–49.
17. Cao, Y., Yao, Z., Sarkar, D., Lawrence, M., Sanchez, G. J., Parker, M. H., MacQuarrie, K. L., Davison, J., Morgan, M. T., Ruzzo, W. L., Gentleman, R. C., & Tapscott, S. J. (2010). Genome-wide MyoD binding in skeletal muscle cells: A potential for broad cellular reprogramming. *Developmental Cell, 18*, 662–674.
18. Mullen, A. C., Orlando, D. A., Newman, J. J., Lovén, J., Kumar, R. M., Bilodeau, S., Reddy, J., Guenther, M. G., DeKoter, R. P., & Young, R. A. (2011). Master transcription factors determine cell-type-specific responses to TGF-β signaling. *Cell, 147*, 565–576.
19. Lund, E., Oldenburg, A. R., Delbarre, E., Freberg, C. T., Duband-Goulet, I., Eskeland, R., Buendia, B., & Collas, P. (2013). Lamin A/C-promoter interactions specify chromatin state-dependent transcription outcomes. *Genome Research, 23*, 1580–1589.
20. Brack, A. S., Conboy, M. J., Roy, S., Lee, M., Kuo, C. J., Keller, C., & Rando, T. A. (2007). Increased Wnt signaling during aging alters muscle stem cell fate and increases fibrosis. *Science, 317*, 807–810.
21. Conboy, I. M., Conboy, M. J., Wagers, A. J., Girma, E. R., Weissman, I. L., & Rando, T. A. (2005). Rejuvenation of aged progenitor cells by exposure to a young systemic environment. *Nature, 433*, 760–764.
22. Mousavi, K., Zare, H., Wang, A. H., & Sartorelli, V. (2012). Polycomb protein Ezh1 promotes RNA polymerase II elongation. *Molecular Cell, 45*, 255–262.
23. Mikkelsen, T. S., Ku, M., Jaffe, D. B., Issac, B., Lieberman, E., Giannoukos, G., Alvarez, P., Brockman, W., Kim, T.-K., Koche, R. P., Lee, W., Mendenhall, E., O'Donovan, A., Presser, A., Russ, C., Xie, X., Meissner, A., Wernig, M., Jaenisch, R., Nusbaum, C., Lander, E. S., & Bernstein, B. E. (2007). Genome-wide maps of chromatin state in pluripotent and lineage-committed cells. *Nature, 448*, 553–560.
24. Ernst, J., & Kellis, M. (2012). ChromHMM: Automating chromatin-state discovery and characterization. *Nature Methods, 9*, 215–216.
25. Scott, L. J., Erdos, M. R., Huyghe, J. R., Welch, R. P., Beck, A. T., Wolford, B. N., Chines, P. S., Didion, J. P., Narisu, N., Stringham, H. M., Taylor, D. L., Jackson, A. U., Vadlamudi, S., Bonnycastle, L. L., Kinnunen, L., Saramies, J., Sundvall, J., Albanus, R. D., Kiseleva, A., Hensley, J., Crawford, G. E., Jiang, H., Wen, X., Watanabe, R. M., Lakka, T. A., Mohlke, K. L., Laakso, M., Tuomilehto, J., Koistinen, H. A., Boehnke, M., Collins, F. S., & Parker, S. C. J. (2016). The genetic regulatory signature of type 2 diabetes in human skeletal muscle. *Nature Communications, 7*, 11764.
26. Braunschweig, U., Hogan, G. J., Pagie, L., & van Steensel, B. (2009). Histone H1 binding is inhibited by histone variant H3.3. *The EMBO Journal, 28*, 3635–3645.
27. Perovanovic, J., DellOrso, S., Gnochi, V. F., Jaiswal, J. K., Sartorelli, V., Vigouroux, C., Mamchaoui, K., Mouly, V., Bonne, G., & Hoffman, E. P. (2016). Laminopathies disrupt epigenomic developmental programs and cell fate. *Science Translational Medicine, 8*, 335ra58.
28. Pickersgill, H., Kalverda, B., de Wit, E., Talhout, W., Fornerod, M., & van Steensel, B. (2006). Characterization of the Drosophila melanogaster genome at the nuclear lamina. *Nature Genetics, 38*, 1005–1014.

29. van Steensel, B., & Henikoff, S. (2000). Identification of in vivo DNA targets of chromatin proteins using tethered dam methyltransferase. *Nature Biotechnology, 18*, 424–428.
30. Vogel, M. J., Peric-Hupkes, D., & van Steensel, B. (2007). Detection of in vivo protein-DNA interactions using DamID in mammalian cells. *Nature Protocols, 2*, 1467–1478.
31. Zhou, V. (2012). *Methods for global characterization of chromatin regulators in human cells*. Cambridge, MA: Harvard University Press. Retrieved November 21, 2014, from http://dash.harvard.edu/handle/1/9414559.
32. van Bemmel, J. G., Filion, G. J., Rosado, A., Talhout, W., de Haas, M., van Welsem, T., van Leeuwen, F., & van Steensel, B. (2013). A network model of the molecular organization of chromatin in Drosophila. *Molecular Cell, 49*, 759–771.
33. Greil, F., Moorman, C., & van Steensel, B. (2006). DamID: Mapping of in vivo protein-genome interactions using tethered DNA adenine methyltransferase. *Methods in Enzymology, 410*, 342–359.
34. Robson, M. I., de las Heras, J. I., Czapiewski, R., Lê Thành, P., Booth, D. G., Kelly, D. A., Webb, S., ARW, K., & Schirmer, E. C. (2016). Tissue-specific gene repositioning by muscle nuclear membrane proteins enhances repression of critical developmental genes during myogenesis. *Molecular Cell, 62*, 834–847.
35. Guelen, L., Pagie, L., Brasset, E., Meuleman, W., Faza, M. B., Talhout, W., Eussen, B. H., de Klein, A., Wessels, L., de Laat, W., & van Steensel, B. (2008). Domain organization of human chromosomes revealed by mapping of nuclear lamina interactions. *Nature, 453*, 948–951.
36. Perovanovic, J. (2015). Nuclear envelope laminopathies: Evidence for developmentally inappropriate nuclear envelope-chromatin associations. *ProQuest Diss Theses*. https://doi.org/10.1186/1756-8935-6-S1-P65.
37. Perovanovic, J., & Hoffman, E. P. (2018). Mechanisms of allelic and clinical heterogeneity of lamin A/C phenotypes. *Physiological Genomics, 50*, 694–704.
38. Boyle, A. P., Song, L., Lee, B.-K., London, D., Keefe, D., Birney, E., Iyer, V. R., Crawford, G. E., & Furey, T. S. (2011). High-resolution genome-wide in vivo footprinting of diverse transcription factors in human cells. *Genome Research, 21*, 456–464.
39. Buenrostro, J. D., Giresi, P. G., Zaba, L. C., Chang, H. Y., & Greenleaf, W. J. (2013). Transposition of native chromatin for fast and sensitive epigenomic profiling of open chromatin, DNA-binding proteins and nucleosome position. *Nature Methods, 10*, 1213–1218.
40. Giresi, P. G., Kim, J., McDaniell, R. M., Iyer, V. R., & Lieb, J. D. (2007). FAIRE (Formaldehyde-Assisted Isolation of Regulatory Elements) isolates active regulatory elements from human chromatin. *Genome Research, 17*, 877–885.
41. Schep, A. N., Buenrostro, J. D., Denny, S. K., Schwartz, K., Sherlock, G., & Greenleaf, W. J. (2015). Structured nucleosome fingerprints enable high-resolution mapping of chromatin architecture within regulatory regions. *Genome Research, 25*, 1757–1770.
42. Loh, K. M., Chen, A., Koh, P. W., Deng, T. Z., Sinha, R., Tsai, J. M., Barkal, A. A., Shen, K. Y., Jain, R., Morganti, R. M., Shyh-Chang, N., Fernhoff, N. B., George, B. M., Wernig, G., Salomon, R. E. A., Chen, Z., Vogel, H., Epstein, J. A., Kundaje, A., Talbot, W. S., Beachy, P. A., Ang, L. T., & Weissman, I. L. (2016). Mapping the pairwise choices leading from pluripotency to human bone, heart, and other mesoderm cell types. *Cell, 166*, 451–467.
43. Wu, J., Huang, B., Chen, H., Yin, Q., Liu, Y., Xiang, Y., Zhang, B., Liu, B., Wang, Q., Xia, W., Li, W., Li, Y., Ma, J., Peng, X., Zheng, H., Ming, J., Zhang, W., Zhang, J., Tian, G., Xu, F., Chang, Z., Na, J., Yang, X., & Xie, W. (2016). The landscape of accessible chromatin in mammalian preimplantation embryos. *Nature, 534*, 652–657.
44. Daugherty, A. C., Yeo, R. W., Buenrostro, J. D., Greenleaf, W. J., Kundaje, A., & Brunet, A. (2017). Chromatin accessibility dynamics reveal novel functional enhancers in C. elegans. *Genome Research, 27*, 2096–2107.
45. Sung, M.-H., Guertin, M. J., Baek, S., & Hager, G. L. (2014). DNase footprint signatures are dictated by factor dynamics and DNA sequence. *Molecular Cell, 56*, 275–285.
46. Sung, M.-H., Baek, S., & Hager, G. L. (2016). Genome-wide footprinting: Ready for prime time? *Nature Methods, 13*, 222–228.

47. Liu, L., Cheung, T. H., Charville, G. W., & Rando, T. A. (2015). Isolation of skeletal muscle stem cells by fluorescence-activated cell sorting. *Nature Protocols, 10*, 1612–1624.
48. Pasut, A., Oleynik, P., & Rudnicki, M. A. (2012). Isolation of muscle stem cells by fluorescence activated cell sorting cytometry. *Methods in Molecular Biology, 798*, 53–64.
49. Schmidl, C., Rendeiro, A. F., Sheffield, N. C., & Bock, C. (2015). ChIPmentation: Fast, robust, low-input ChIP-seq for histones and transcription factors. *Nature Methods, 12*, 963–965.
50. Sherwood, R. I., Hashimoto, T., O'Donnell, C. W., Lewis, S., Barkal, A. A., van Hoff, J. P., Karun, V., Jaakkola, T., & Gifford, D. K. (2014). Discovery of directional and nondirectional pioneer transcription factors by modeling DNase profile magnitude and shape. *Nature Biotechnology, 32*, 171–178.
51. Meyer, C. A., & Liu, X. S. (2014). Identifying and mitigating bias in next-generation sequencing methods for chromatin biology. *Nature Reviews Genetics, 15*, 709–721.
52. Askjaer, P., Ercan, S., & Meister, P. (2014). Modern techniques for the analysis of chromatin and nuclear organization in C. elegans. *WormBook*. https://doi.org/10.1895/wormbook.1.169.1.
53. Nègre, N., Hennetin, J., Sun, L. V., Lavrov, S., Bellis, M., White, K. P., & Cavalli, G. (2006). Chromosomal distribution of PcG proteins during Drosophila development. *PLoS Biology, 4*, e170.
54. Clark, S. J., Lee, H. J., Smallwood, S. A., Kelsey, G., & Reik, W. (2016). Single-cell epigenomics: Powerful new methods for understanding gene regulation and cell identity. *Genome Biology, 17*, 72.
55. Cusanovich, D. A., Daza, R., Adey, A., Pliner, H. A., Christiansen, L., Gunderson, K. L., Steemers, F. J., Trapnell, C., & Shendure, J. (2015). Multiplex single-cell profiling of chromatin accessibility by combinatorial cellular indexing. *Science, 348*, 910–914.
56. Pliner, H., Packer, J., McFaline-Figueroa, J., Cusanovich, D., Daza, R., Srivatsan, S., Qiu, X., Jackson, D., Minkina, A., Adey, A., Steemers, F., Shendure, J., & Trapnell, C. (2017). Chromatin accessibility dynamics of myogenesis at single cell resolution. *bioRxiv*. https://doi.org/10.1101/155473.
57. Buenrostro, J. D., Wu, B., Litzenburger, U. M., Ruff, D., Gonzales, M. L., Snyder, M. P., Chang, H. Y., & Greenleaf, W. J. (2015). Single-cell chromatin accessibility reveals principles of regulatory variation. *Nature, 523*, 486–490.
58. Davie, K., Janssens, J., Koldere, D., Pech, U., Aibar, S., De Waegeneer, M., Makhzami, S., Christiaens, V., Gonzalez-Blas, C. B., Hulselmans, G., Spanier, K., Moerman, T., Vanspauwen, B., Lammertyn, J., Thienpont, B., Liu, S., Verstreken, P., & Aerts, S. (2017). A single-cell catalogue of regulatory states in the ageing Drosophila brain. *bioRxiv*. https://doi.org/10.1101/237420.
59. Rotem, A., Ram, O., Shoresh, N., Sperling, R. A., Goren, A., Weitz, D. A., & Bernstein, B. E. (2015). Single-cell ChIP-seq reveals cell subpopulations defined by chromatin state. *Nature Biotechnology, 33*, 1165–1172.
60. Kind, J., Pagie, L., Ortabozkoyun, H., Boyle, S., de Vries, S. S., Janssen, H., Amendola, M., Nolen, L. D., Bickmore, W. A., & van Steensel, B. (2013). Single-cell dynamics of genome-nuclear lamina interactions. *Cell, 153*, 178–192.
61. Roadmap Epigenomics Project – Home [Online]. (n.d.). Retrieved from January 1, 2018, from http://www.roadmapepigenomics.org/.
62. Wang, J., & Song, Y. (2017). Single cell sequencing: A distinct new field. *Clinical and Translational Medicine, 6*, 10.

Chapter 4
Guidelines for Bioinformatics and the Statistical Analysis of Omic Data

Surajit Bhattacharya and Heather Gordish-Dressman

4.1 Bioinformatics: An overview

The Human Genome Project [1] was initiated to find all the nucleotides that constitute a human genome and identify and map genes making up that genome. The Human Genome Project, along with the other model organism projects, have enabled us to better understand the genetic and molecular underpinnings of developmental stages and disease conditions. These projects have also produced petabytes of data, mostly sequences of nucleotides, which are not comprehensible to a biologist until and unless it is given a biological perspective. Bioinformatics is a multidisciplinary branch of science, utilizing knowledge from many other fields including physics, mathematics, computational science, and statistics, to assist in transforming these raw nucleotide sequences to a more comprehensible biological dataset (Fig. 4.1). Bioinformatics can be broadly classified into two groups, genomics and proteomics. Genomics includes the study of the genomic data pertaining to defects in transcriptional and posttranscriptional mechanisms, while proteomics pertains to changes in proteins occurring mostly during the translational and post-translational phase. In this section we discuss the bioinformatics tools and algorithms used in various kinds of genomic experiments. Bioinformatics for transcriptome analyses can be found in the transcriptome chapter.

Genomic analyses focus on deoxyribonucleic acid (DNA), which is the building block of any biological system and is the primary component of the central dogma.

S. Bhattacharya
Center for Genetic Medicine Research, Children's National Medical Center, Washington, DC, USA

H. Gordish-Dressman (✉)
Center for Translational Research, Children's National Medical Center, Washington, DC, USA

Department of Pediatrics, The George Washington University School of Medicine and Health Sciences, Washington, DC, USA
e-mail: HGordish@childrensnational.org

J. G. Burniston, Y.-W. Chen (eds.), *Omics Approaches to Understanding Muscle Biology*, Methods in Physiology, https://doi.org/10.1007/978-1-4939-9802-9_4

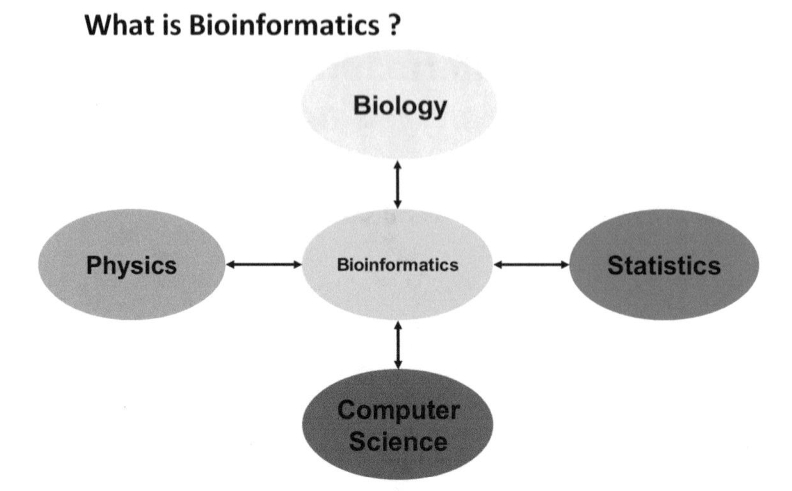

Fig. 4.1 Bioinformatics is a multidisciplinary branch of science: bioinformatics is a multidisciplinary branch of science, utilizing ideas from major branches of science such as physics, statistics, biology, and computer science

DNA is transcribed into ribonucleic acid (RNA), and messenger RNA (mRNA) directs protein translation. DNA is found in the nucleus and is composed of four major groups of nitrogen-based organic compounds (nucleotides): deoxyadenosine (A), deoxyguanosine (G), deoxythymidine (T), and deoxycytidine (C). These four nucleotides are arranged in different combinations to form single-stranded DNA. As A forms a chemical bond with T and C bonds with G, the DNA arranges itself into the shape of a double-stranded helical structure where each strand complements the other. If one strand has a nucleotide arrangement of ATTTCGATA, the second complementary strand will have an arrangement of TAAAGCTAT. This arrangement of nucleotides is called a sequence and the two strands of DNA sequences are referred to as the forward strand (read from left to right; also called 5′–3′) or the reverse strand (read from right to left; also called 3′–5′). RNA, another common biological material studied in genomics, differs slightly from DNA. Besides being made up of ribonucleic acid rather than deoxyribonucleic acid, RNA contains a unique nucleotide, uracil (U), in place of thymine.

From its isolation from pus in late 1868 [2] to the end of the Human Genome Project, the identification of the sequence of DNA has been an important aspect in understanding the functionality of the genes. This chapter is intended to be a resource for researchers interested in designing omic experiments or analyzing data from such experiments. It is organized into two major applications. The first has a focus on bioinformatics tools and techniques, and the second has a focus on statistical analyses. This chapter is intended to be a high-level introduction for researchers interested in performing their own analyses. It describes the major components of bioinformatics and statistical analysis and also directs the reader to key tools and sources of further information. Unfortunately it is beyond the scope of this chapter to provide a comprehensive discussion of statistical theory and complex statistical models. This chapter discusses the general study design and aspects that should be taken into account before an experiment is begun. We describe some basic principles of statistical analysis and some commonly used methods, but further guidance from a statistician will be required for more complex study designs and statistical models.

4.2 Major Applications

4.2.1 Bioinformatics of Genomics Data

The first section discusses the bioinformatics tools and algorithms used in genomics. It describes typical workflows and the tools available for performing an omic experiment and underscores the importance of both the tools being used and a clear understanding of the underlying algorithm. One of the primary goals of any genomic research is to understand whether there are any changes or variants in the genome of a subject as compared to a reference genome. The reference genome is a representative genome of an organism, including humans, that is made up of multiple samples and is a representative example of a species' set of genes. The variation from this reference genome can be of two types: single nucleotide polymorphisms (SNP) and structural variants (SV). While SNPs constitute a change in a single nucleotide, SVs constitute a variation of $\geq$50 kilobases (KB) which can be an insertion, deletion, inversion, translocation, or duplication. It is these variants, or differences, which we are interested in when we perform omic experiments.

Next-generation sequencing (NGS) is the common term used for the modern-day high-throughput sequencing techniques, which allow one to define the precise order of nucleotides in the DNA of the organism sequenced. As this chapter focuses on the bioinformatics aspect of omic experiments, we will focus more on the results obtained from sequencing instruments than the sequencing techniques used. A general overview of sequencing techniques can be found in the following reviews [3, 4]. Here we briefly discuss the workflow for whole genome sequencing (WGS) analysis, obtained from a short-read sequencer, such as that produced by Illumina. A similar pipeline is used for exome sequencing. The pipeline can be divided into five parts (workflow is shown in Fig. 4.2):

1. *Preprocessing of raw sequences*: Raw output from sequencers, referred to as binary base calls (BCL), is converted to the human readable Fastq format [5]. Fastq format generally has three lines representing the sequence header, the sequence, and a quality score (*PHRED* score) [6, 7]. The PHRED score is a measure of the quality of the nucleotide identification associated with each base written in ASCII format. The PHRED score Q is defined as

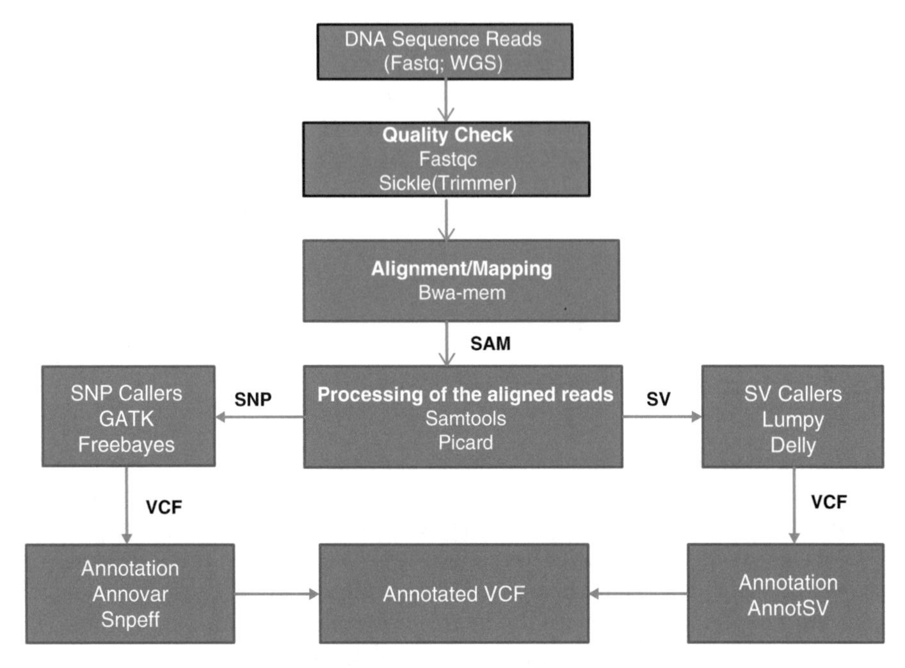

Fig. 4.2 Workflow for WGS: sequence files obtained from sequencers are converted to Fastq files, and an initial quality check using *FastQC* is performed. If the quality is low, the file can be trimmed using a trimming software such as *Sickle*. Next, alignment to the reference genome is performed using *BWA*. Aligned SAM files are processed using *Samtools* and *Picard* based on the type of variants to be identified. For single nucleotide polymorphisms (SNPs) and small insertion deletions (indels), *GATK* or *Freebayes* is chosen. For larger structural variants (SV), tools like *Delly* and Lumpy can be used. The output of all of these tools is a variant call format (VCF) file. An annotation tool for annotating the VCF file is chosen based on the variant that needs to be identified. *Annovar* and *Snpeff* are used for SNP annotation and *AnnotSV* for SV annotation. The final product is an annotated VCF file

$$Q = -10\log P$$

where P is the base-calling error probability. Thus, a larger Q value indicates a greater accuracy in identifying the base. The generally accepted PHRED score threshold is 30 which translates to the chances that the base was incorrectly identified as 1 out of 1000. The tool *FastQC* [8] can be used not only to provide a visual representation of PHRED score per sequence but to provide additional useful information such as sequence GC content, sequence length distribution, and sequence duplication distribution. If a sequence is found to be of low quality, trimming tools such as *cutadapt* [9], *sickle* [10], and *scythe* (tool available at https://github.com/vsbuffalo/scythe) can be used to remove unwanted sequences. As one example, sickle allows one to input a PHRED score threshold so that the user can define which sequences will be trimmed due to quality issues. Each of these tools has various capabilities and the documentation for each should be consulted and understood before using.

2. *Alignment of the sequence*: Multiple tools/algorithms are available for the alignment/mapping of the whole genome sequence(s) to the reference genome. There are several important aspects to consider when choosing an alignment algorithm/software. These include the accuracy of alignment to the reference genome, the amount of computational memory used, and the time required for the alignment. One of the most important and commonly used algorithms is the Burrows-Wheeler algorithm (*BWA*) [11]. The BWA uses backward search with a compression algorithm, the Burrows-Wheeler transform (BWT) [12], to align sequence reads to the reference genome at a reduced memory usage and time [13]. The BWA software package consists of three algorithms, BWA-backtrack, BWA-SW, and BWA-MEM. BWA-backtrack is used for sequence reads that are shorter than 70 base pairs, while both BWA-SW and BWA-MEM are used for longer reads. The choice between BWA-SW and BWA-MEM is dependent on the platform used to generate the reads and the characteristics of the data. This algorithm can align 7 Gbp (giga base pairs) to the human reference genome per CPU day [13]. Output from this alignment step is a Sequence Alignment Map (SAM).
3. *Processing of the aligned files*: The output from this step needs to be further processed before it can be used for calling variants. Read group names are added to the aligned reads using *Picard* (https://broadinstitute.github.io/picard/) to identify each of the reads from different runs separately. *Samtools* [14] uses a merge function to combine different SAM files into a single merged binary aligned BAM file. Quality control processes can then be applied to the merged files using Samtools, and PCR duplicates can be removed by Picard Mark Duplicates.
4. *Identifying single nucleotide polymorphisms and structural variants*: Typically, short-read sequencers are more adept at identifying or calling small nucleotide variants and small insertion/deletions (indels) than structural variants. The Genome Analysis Toolkit (*GATK*) [15] is the standard and most commonly used pipeline for the identification of small nucleotide polymorphism and small indels from an aligned and processed BAM file. Other softwares, *Freebayes* [16] and *Heap* [17], are also available with Heap having the advantage in the detection of SNPs at low coverage.

   Although short-read sequencers are not ideal for identifying structural variants, there are multiple tools that can detect and identify structural variants from BAMs obtained by aligning short-read sequencer reads. All SV calling tools for short reads are based on five basic ideas:

   (a) *Read pair (RP)*: Discovers SVs by looking at the span and orientation of paired end reads. This method is used to identify almost all types of SVs. Pairs that are mapped far apart may denote deletion, whereas closer read pairs denote insertion. Orientation inconsistencies denote inversion or tandem duplications [18]. Some of the tools associated to this method are *BreakDancer* [19], *PEMer* [20], and *MODIL* [21].

(b) *Read count (RC)*: This method can be also known as *read depth*. In these algorithms, a Poisson or modified Poisson distribution is assumed for the mapping depth, and deviation from that distribution denotes deletion or duplication [22]. This kind of SV is also called copy number variants (CNV). Tools associated with this method are *EXCAVATOR* [23], *CNVnator* [24], and *BIC-seq* [24].

(c) *Split read (SR)*: This method defined SV break points by identifying deviations between the sample sequence and the reference genome. A gap in the sample sequence denotes deletion, while additional sequence (above that in the reference) denotes insertion. Tools include *Socrates* [25] and *Splitread* [26].

(d) *De novo assembly (AS)*: In these group of algorithms, the short fragments are reassembled to form the original sequence, to identify the SV [27]. Tools include *Cortex* [28] and *Magnolya* [29].

(e) *Hybrid methods*: These algorithms use a combination of two or more methods discussed above. For example, *Delly* [30] and *Lumpy* [31] use read pair and split read algorithms, to detect SV.

As there are multiple methods to detect SVs, it is advisable to get a consensus result from multiple methods to confidently detect a true set of SVs in a sample. Tools like *svMerge* [32], *SURVIVOR* [33], and *Parliament* [34] combine results from multiple SV-identifying tools, to get a consensus output in variant call format (VCF).

5. *Annotation*: The output from the SV identifiers are in the form of a VCF, which includes information on the chromosome number and location of the variant along with type of SNP/SV and quality. It does not, however, contain information on the gene affected or whether or not the variant is a rare variant. This information is added by annotation tools. Annotation tools like *Annovar* [35] can perform annotation for both SVs and SNPs. Using Gene transfer format (GTF), Annovar identifies genes affected and uses databases such as *dbSNP* [36] (for SNPs) and Database of Genomic Variants [37] (*DGV*; for SVs) to identify the frequency of occurrence of the particular SV across different populations. *Clinvar* [38] database is used to identify the clinical significance of the variants (whether it causes disease or not), whereas the *CADD* [39, 40] tool is used to classify the variant based on its pathogenicity. *Snpeff* [41] and *SnpSift* [42] can be used to annotate and filter SNPs and small insertion deletion (indels; <50 bases), respectively, whereas *AnnotSV* [43] performs the same task for SVs.

4.2.2 Other Bioinformatics Analyses

Beyond expression analysis and variant discovery, NGS and microarrays can also be used to determine and evaluate the effect of other biological factors that affect a system. They include:

1. *Transcription factor binding site*: Chromatin immunoprecipitation (*ChIP*) is an immunoprecipitation technique by which the genomic region to which a transcription factor binds is extracted and then analyzed using either microarray or sequencing techniques. The basic working of microarray and sequencing remains the same for ChIP experiments, only the analysis tool pipeline changes.

   (a) *Microarray*: These experiments are known as *ChIP-on-chip*. The design of the experiments is similar to gene expression determination experiments, where the control is without the antibody to pull down the genomic region, while the experimental condition has the pulled down genomic region. Tiling Analysis Software (open-source software from Affymetrix) is used to process raw data into signals, with associated p-values for each probe. This is followed by visualization using Integrated Genome Browser (IGB), a proprietary software from BioViz. Another proprietary software, Partek, can also be used to analyze and visualize the regions to which transcription factors bind to. An open-source Bioconductor R pipeline is also available [87] which uses *Ringo* [88] for processing the raw microarray data into normalized data, *BioMart* [89] to annotate the data, and *topGO* [90] to perform the gene ontology. Ringo also has functions to visualize the genomic region.

   (b) *RNA-seq*: These experiments are known as *ChIP-seq*. Preprocessing, alignment, and processing of BAMs are done in the same fashion as RNA-seq data. FastQC is used for quality check, followed by alignment with Bowtie or BWA, followed by processing of SAM files by Samtools and Picard. To identify genomic regions to which transcription factors bind, we identify regions in the genome where maximum reads map to. These are called peaks. Peak calling is performed using tools like model-based analysis of ChIP-Seq (*MACS*) [91] and spatial clustering for identification of ChIP-enriched regions (*SICER*) [92]. Normalization and fold change calculation of transcription factor binding between two samples (control and experimental) are estimated using *edgeR*, *Gfold*, and *Deseq2*.

2. *Methylation*: Determination of methylation is done using bisulfite treatment. This converts unmethylated cytosine to uracil but leaves methylated cytosine unchanged [93]. Microarray and/or sequencing techniques can be then performed to determine the difference in methylation between an experimental condition and a control condition.

   (a) *Microarray*: The most popular platforms are Illumina 450 K and Epic (850 K) arrays. These arrays are named after the number of probes used; the 450 K has 450,000 probes of known methylation region and the 850 K is composed of

850,000 probes. There are multiple tools to analyze this data. Partek, a proprietary software, performs differential methylation and visualization of methylation regions. Open-source workflow *Champ* [94, 95], a R Bioconductor workflow, uses other Bioconductor packages like *minfi* [96] for analysis and visualization of methylation, a modified version of *combat* function from sva package [97] for batch correction, *limma* for differential expression calculation, and *plotly* [98] for visualization.

(b) *Sequencing*: The workflow can be divided into two steps.

   (i) *Preprocessing and alignment-based identification*: The preprocessing and the alignment steps are similar to preprocessing step as described in the Genomics and Transcriptomics sections; the only differences are in the tools used for alignment. FastQC is used for quality check and sickle used for trimming. Most of the alignment methods use a modified version of other short read alignment algorithms like Bowtie2. Tools like *Bismark* [99], *BS-Seeker* [100], and *B-Solana* [101] use Bowtie algorithm, with methods to handle cytosine to thymine (uracil in RNA) conversion. LAST [102], another algorithm, uses score matrix to handle C-T conversion, and *Bisulfighter* [103] uses this algorithm for alignment. Another algorithm, BSMAP [104], converts the thymine to cytosine in the bisulfite reads in silico, in positions in which cytosine is present in the reference.

   (ii) *Differential methylation*: The best tool to use is based on whether replicates are present in the experiment to be analyzed. For experiments in which there are no replicates available for control or experimental conditions, methylkit [102] or RnBeads [105] uses Fisher's exact tests, and *ComMet* [103, 106] (part of *Bisulfighter* methylation analysis pipeline) or *Methpipe* [107] uses hidden Markov models. For experiments having replicates, tools like *limma*, *BSmooth* [108], and *Biseq* [109] use regression models. Good reviews of these tools and methods can be found in [110, 111].

4.2.3 Platforms Used

Most of the pipelines used in the discussion involve a combination of a Linux environment and R IDEs like Rstudio. Open-source user interface (UI) option for analyzing NGS studies is limited, with Galaxy [116] being the only reliable option. Cloud platform UIs like DNAnexus and artificial intelligence RNA-seq (AIR) are proprietary and are subscription based. Cloud-based platforms such as Amazon and Microsoft Azure can also be used for processing large amounts of genomic data, but they are mostly command line based.

4.2.4 Downstream Analysis and Visualization

After the analysis to identify the most significantly expressed genes is complete, multiple downstream analyses can then be performed to give biological perspective to the genes defined. Following are the two most important downstream analyses done on transcriptomic analyzed data:

1. *Gene ontology and pathway analysis*: To better understand the biological system studied in a given experiment, one needs to understand the function and pathways of the genes that are expressed in the system. Gene ontology [117] (*GO*) classification tools can classify the function of the genes under three broad classes: cellular components (*CC*), molecular functions (*MF*), and biological processes (*BP*). This kind of analysis is called a gene enrichment analysis (GEA). GEA can also be divided into three categories:

   (a) *Singular enrichment analysis (SEA)*: A user-defined gene list derived from any high-throughput NGS/microarray experiment can be used as an input [118]. Generally, the input genes are those highly significant genes based on user-defined arbitrary threshold (e.g., multiple testing adjusted p-value $<$ 0.05). The genes are classified as falling into the three broad categories (BP, CC, and MF) described above. Fisher's exact test [119] and its modifications (*EASE* score [118]) or a chi-square test [120] is used in this method to determine whether the genes are related to the categorization based on function. Tools such as *DAVID* [118, 121], *GOStat* [122], and *Bingo* [123] perform singular enrichment analysis.
   (b) *Gene set enrichment analysis (GSEA)*: In gene set enrichment analysis, all genes from a high-throughput genomic experiment are used as an input, thus ensuring the analysis is free from any potential bias due to arbitrary thresholds such as those used in SEA [124]. This allows even genes with small differential expression changes to be considered for further enrichment analysis. Maximum enrichment scores (MESs) are calculated from the rank order of all gene members in an annotation category. These maximum enrichment scores are analyzed and p-values calculated using either Kolmogorov-Smirnov-like statistics to compare the observed MESs to those obtained with from a random distribution or with parametric statistics which compare fold changes between experiments. *ErmineJ* [125] and *FatiScan* [126] can be used to perform this type of analysis.
   (c) *Modular enrichment analysis (MEA)*: Modular enrichment analysis combines SEA-kind enrichment analysis with network discovery algorithms, to allow term to term relationships. Kappa statistics of agreement are used for this analysis. Due to the structure of the kappa statistics, genes which do not occur in multiple neighboring terms are excluded from the analysis. Some tools that perform this type of analysis are *ADGO* [127], *DAVID*, and *GeneCodis* [128].

These gene ontology classification tools combine data from multiple sources including *KEGG* [129] for pathway analysis, *Pfam* [130] for protein domains, and *TRANSFAC* [131] for gene regulation.

2. *Correlation networks*: Another important piece of information from the obtained gene lists are whether or not there is any relationship between the statistically significant gene lists. Although we learn a lot about genes that share common functionality from the GEA, tools like *GeneMania* [132] provide the user with information on the interaction between genes, thus yielding a more comprehensive picture of the interaction pathways. GeneMania provides the users with predictions of co-expression, co-localization, and physical interactions between genes. Biological General Repository for Interaction Datasets (*Biogrid*) [133] is another database which provides the user with chemical, genetic, and protein interactions using known experimental results and is curated regularly. *STRING* [133], a protein interaction database, is used to understand the interaction between the proteins translated from the genes. One can also use weighted gene correlation network analysis (WGCNA) [134], a package in R [135], which uses expression data from microarray or RNA-seq to construct a correlation network between genes in a given experiment.

Visualization Tools like *FastQC and MultiQC* [136] are used to evaluate and visualize the quality of the sequencing from both Fastq and BAM data. Visualization of expression data can be done using Heatmap.2 function of the *gplots* [137] package in R. *GOPlot* [138], an R package, which can be used to visualize Gene Ontology data.

4.2.5 Conclusion

Although newer long-read sequencing techniques as well as optical mapping techniques are being developed to better understand variation in genes and changes in expression, no technology has yet displaced short-read sequencing and/or microarrays. With the advancement in computational hardware, as well as development of more extensive computational algorithms to better handle big data, bioinformatics tools for analysis of NGS data are rapidly changing. In these changing times, one should not only concentrate on the tools being used, but also better understand and appreciate the underlying algorithm to better evaluate his/her data. In addition, a thorough understanding of experimental design and statistical analysis is incredibly important to yield a thoughtful and meaningful analysis.

4.3 Statistical Analysis

4.3.1 Study Design Principles

Study design at its most basic level is defined as the methods for planning the collection of data in order to obtain the maximum amount of information using the least amount of resources. As such, good study design is an integral part of any experiment. There are many available resources that discuss study design in general, and an Internet search of study design fundamentals will yield hundreds of sources. However, omic experiments have certain common traits that make many typical study design elements less applicable. We often do not have large numbers of samples available or the high cost of the technologies limits the number of samples we have the resources for. We are often testing many more parameters than we have samples for, causing difficulties with our statistical analyses [139, 140]. Lastly, due to the scarcity of individual samples, we are often tempted to test complicated hypotheses that we do not have the data to truly support.

Before we begin an experiment, there are several things we should consider so that our experiment can be planned to appropriately answer our scientific question. We need to have a testable hypothesis, we need to define our replicates (how many we have and what type they are), we need to understand the sources of variability in our sample and how it relates to the population, and we need to verify that our sample size is adequate to answer our question. Lastly, we need to understand what our collected data points are representing.

4.3.1.1 Hypothesis

Having a testable hypothesis may seem an obvious step; however, not all hypotheses are well thought out or testable. Omic experiments are often used for hypothesis "generation" where we use the data we have to inform what is happening in a biological system. We then use this information to generate hypotheses that can be tested in future experiments. Because of this feature, it is mistakenly thought that omic experiments are discovery in nature and do not need an a priori defined hypothesis. While this is not entirely false, it is also not true [141]. One may have a more general question to ask such as "are RNA levels related to biceps strength?" with the intention of discovering what RNA levels change in response to strength. However, when we perform our statistical analysis, we are in fact testing a hypothesis. That hypothesis may involve assessing the relationship between RNA levels at one particular strength level, or assessing the change in RNA levels as strength changes, or assessing whether or not the relationship between RNA level and strength is different with respect to some other factors such as gender. These three hypotheses are all tested using different statistical methods and require different study designs and numbers of samples. So, we must know what hypothesis we want to test before we plan our experiment, even if our goal is only to "discover" what

RNA levels are important to us. If we don't consider a hypothesis, we may be left at the end with data that cannot answer our questions.

Imagine a case where we are interested in assessing whether exercise in men has an effect on the expression level of gene *X*. We measure gene expression in 10 men who exercise on a regular basis and in 10 men who are sedentary. If we are truly interested in whether or not exercise causes a change in expression of gene *X*, we cannot test that hypothesis with this sample. All we have from our sample design is expression levels in exercisers and expression levels in sedentary individuals. We may find significant differences in the expression of gene *X*; however, we cannot say those differences are due to exercise. They could be due to any number of differences between the two groups of men, not necessarily related to exercise. Our intentions may have been to test the hypothesis that exercise induces changes in gene expression, but our sample cannot test that hypothesis. In order to test our intended hypothesis, we would need repeated measurements of expression on the same individuals both before and after exercise. Only then could we infer that the exercise was, in part, responsible for any differences observed in expression levels.

The above example shows how it can be a mistake to design an experiment without a clear hypothesis in mind. We can spend valuable resources collecting data that does not, in the end, meet our needs.

4.3.1.2 Replicates

The understanding and choice of replicates is often a source of confusion because the term "replicates" has various definitions depending on who is defining it. At its simplest, a replicate is a measurement taken more than once. Some refer to multiple measurements taken from the same sample at the same time as replicates. Others refer to measurements taken from the same sample at different times as replicates. And yet others consider replicates to be different aliquots of the same RNA extraction or multiple arrays hybridized with the same RNA.

We can divide replicates into two types, biological and technical, which serve different purposes. Technical replicates are used to establish the variability, or experimental error, of the measurement technique. They are performed on the same biological sample so that the differences we see can be wholly attributed to the measurement technique. All measurement techniques have variability in how well they measure the parameter of interest and we need to quantify this to better allow us to measure other sources of variability we encounter. Those performing omic experiments should be familiar with this phenomenon as we must know how much noise is in the measurement to accurately assess the signal. Biological replicates, on the other hand, are broadly speaking biologically different samples. They are used to establish the biological variability which exists between organisms which should be identical, and they are typically the reason we are performing the experiment in the first place. The biological entities in our experiment represent the wider population. We can only make inferences, by applying inferential statistics, about the population from which our samples were taken. When we apply statistical tests to our sample we are inferring what we would expect to see in the larger population. Vaux et al. [142] provides an excellent overview of the difference between technical and biological replicates and describe why one can be used in hypothesis tests and inferential statistics and the other cannot. A particularly apt description of the mistake in the blurring of technical and biological replicates is given by Bell [143].

Consider the three following experiments, each of which contains replicates. The first experiment assesses gene expression in 10 separate mice. Here we have 10 biological replicates and we expect expression to be the same in each mouse. We are measuring how much difference in expression exists in our mice. This sample of 10 mice is representative of the population of all mice and we can apply inferential statistics to infer that our conclusions apply to the population. The second experiment assesses protein expression in a cell line that has been aliquoted into 20 plate wells. Here we have 20 technical replicates; we expect protein levels to be exactly the same from the same cell line; therefore, we are measuring the variability in the measurement method. Our population here is this single cell line. The third experiment is more complicated. We have one cell line that is aliquoted into two petri dishes, one of which is treated with a potential drug. From each dish, we take three aliquots and measure gene expression of gene Y. Here we have three technical replicates that could be used to estimate the variability in the gene expression measurement, but we do not have any biological replicates; all measurements were made on the same sample. We may be tempted to compare gene expression between the treated samples and the untreated samples and attribute any differences we see to the treatment. However, our population remains one single cell line. Any conclusion we make from that comparison applies only to that single cell line; we cannot infer the same would be true in other cell lines.

We must be careful to understand what our replicates actually represent. Because the p-values calculated from statistical tests are based in part on the number of data points we have, we can inadvertently artificially increase our sample size and bias our conclusions by mistaking technical replicates for biological replicates.

4.3.1.3 Understanding Sources of Variability

All statistical tests assume that what we observe in our sample can be inferred in the population under study. When we understand our sample, we can be more confident that our conclusions are generalizable to the population under study. We can appropriately deal with the many sources of variability, both biological and technical, that, when ignored, can severely impact our statistical analysis and the conclusions from that analysis.

Many sources of variability are familiar to those performing omic experiments. These can include date and method of data collection or preparation, protocol used, the technician performing preparatory analyses, software used for data processing, and a host of other experiment-/platform-specific characteristics. We can account for these potential differences in our samples either statistically or technically; however, we need to have this information to do so. We can plan our experiment so that all samples use the same preparation techniques and are performed by the same technician, or we can account for these differences through the use of covariates in our statistical tests.

Equally important are sources of variability within our study population. These may not be as familiar to those performing omic experiments but they can affect the conclusions from our experiments. When one is building a hypothesis based on a defining variable, such as the effect of a drug, both the samples receiving the drug and those not receiving the drug should be as similar to each other as possible. This becomes more difficult when the defining variable is also an inherent difference between samples, for example, assessing gene expression in individuals with a disease compared to those without. It is difficult to find affected and unaffected individuals who are similar in all other respects, making any expression differences possibly due to other factors rather than the disease. The more completely we characterize and understand the variability in our samples, the more confidence we have in the validity of our conclusions.

4.3.1.4 Adequate Sample Size

A last area of importance in study design, but one that is often the most difficult to deal with, is ensuring we have an adequate sample size for the hypothesis we are testing. All of the previously described areas of study design are important in determining what your sample size should be or what power you have given an already existing sample. Sometimes we have all of the information we need to adequately perform our sample size or power calculations; however, more often we do not. This section does not aim to discuss the theory behind sample size or power calculations or to fully describe all of the individual techniques that have been developed for specific types of experiments. It is instead intended to make the reader aware of what is necessary to complete these calculations and to impart their importance. A more thorough and easily readable discussion of sample size calculations can be found in Whitley and Ball [144]. An additional resource by Billoir et al. [145], although more technical, discusses sample size and power specifically for high-throughput experiments with a particular focus on metabolomics.

Sample size and power calculations utilize the same mathematical equations and have common elements. A checklist for the elements needed for power and/or sample size calculations include:

1. Well-defined hypothesis
2. The statistical test we will use to test our hypothesis
3. The effect size we expect to see (i.e., how much difference are we trying to detect)
4. How much variation in our outcome can we expect
5. The significance level (type I error rate) we want to test the hypothesis at
6. The power (type II error) we want/have
7. The sample size we need/have

You'll notice that the list includes both power and sample size. If one wants to determine sample size, power must be provided. If one wants to determine power, sample size must be given. One can choose to calculate the power that a sample has to test the hypothesis or calculate the number of samples needed to test the hypothesis at a given power level. We'll discuss what each of these components is and how to define them.

A well-defined hypothesis and the statistical test we will use to test it are necessary because each statistical test has its own equation to calculate sample size and/or power. The equation to calculate sample size for a *t*-test is very different than that for a chi-square test. We must know what test we will perform so that we can use the correct equation. The choice of statistical test, as described above, is determined by the hypothesis one is testing. So, our first step is to define our hypothesis and determine what test will be used. Only then can we perform the appropriate analysis to define our sample size/power.

The effect size describes how large of an effect you would define as statistically significant. In other words, how large of a difference in your dependent variable would you expect to see or define as meaningful. If you have preliminary data available, this difference may come from the difference you've observed in that data. Or you may decide that only a difference of a certain amount is meaningful; therefore, you want to have a large enough sample to detect that effect size. Imagine you are planning an experiment to test the hypothesis that expression of gene *A* is different in men and women. You may have some previous data that indicates expression in men is 10% greater than in women. This is the effect size, or difference, you would like to plan your experiment to detect. Or you may not have any previous data but consider any difference between men and women smaller than 10% to be meaningless; again 10% would be the effect size you are planning to detect. Defining the effect size can be difficult, especially if you have no preliminary data nor an idea of what would be meaningful. If this is the case, we have to improvise and make our best guess as to what the effect size will be and understand that detecting a smaller effect size requires a larger sample size. We can use literature sources of similar experiments as a guide, or we can calculate sample size or power for a range of effect sizes and hope that the effect size that ultimately exists in our sample falls within that range.

The expected variability is directly related to our effect size. It is an estimate of how much variability we expect to find in our dependent variable. Not surprisingly, the more variable our dependent variable is, the larger the sample size needed will be. Unless we have preliminary data, the expected variability can be difficult to estimate. Like effect size, we can make a best guess based on similarities in literature sources or use a range of values.

The significance level, or the type I error rate, is the level at which we want to test our hypothesis. It defines the false-positive rate that we are willing to accept and still define our conclusion as statistically significant. This value is conventionally set at 0.05 for any single statistical test; 5% is the level of uncertainty we are comfortable with. However, in omic experiments, we are rarely performing only a single statistical test and are often performing many hundreds or thousands of tests. To account for this, we often adjust the significance level in our sample size/power calculations to account for these many tests. This phenomenon is described in more detail in the multiple testing section of this chapter.

The last elements are power and sample size. These are the values we are interested in calculating. If we want to determine what sample size is needed to test our hypothesis, we need to define the power we are comfortable with. Power, or type II error, is the rate at which we expect to conclude there is a statistically significant difference when there truly is. It is defined as 1 minus the false-negative rate (1-FN). Conventionally we perform statistical tests with 80% power; however, we are free to increase our power if warranted. Once we define these elements, we have all of the necessary information to calculate the number of samples we need to test our hypothesis at the given significance and power levels.

Sometimes we already have our sample in hand. It may be a sample of convenience that happens to be available, or we may know that we have the money and resources for a certain number of samples. In these cases, we can calculate how much power we have given our sample. This is an important step that is not only usually required in funding proposals, but one that is in our best interest to perform. Underpowered studies, those that do not have adequate power to test their hypotheses, are a common problem and often result in inconclusive results. At the end of an experiment, a nonsignificant *p*-value indicates that either there truly is no effect or the sample was not powered to detect the effect. If you know that your analysis was adequately powered, you can be confident in concluding there is no effect. If on the other hand you have an underpowered study, you're left with an inconclusive answer. Assuring that your experiment is adequately powered allows one to adequately answer the question they are interested in and assures the funding agency that the money spent on the experiment will be of use.

The table below (from [144]) shows the relationships between all of the elements that go into sample size or power calculations. As mentioned above, a smaller effect size or a dependent variable with a large level of variability requires a larger sample size. An experiment with greater power requires a larger sample size than one with lower power.

Factor	Magnitude	Impact of identification of effect	Required sample size
P-value	Small	Stringent criteria; may be difficult to achieve significance	Large
	Large	Relaxed criteria; easier to attain significance	Small
Power	Low	Identification unlikely	Small
	High	Identification more probable	Large
Effect size	Small	Difficult to identify	Large
	Large	Easy to identify	Small
Variability	Small	Easy to identify	Small
	Large	Difficult to identify	Large

4.3.1.5 Data Representation

One last topic the researcher should be aware of is the pre-data processing that has occurred prior to statistical analysis. Just as it's important to understand what hypothesis you are testing and what your replicates represent, it's also important to understand what one's data represent. Typically, omic experiments utilize data from an instrument that goes through various steps of processing and normalization before any statistical analysis is performed. Rarely is raw data collected from an instrument used in analysis. Each type of experiment, i.e., proteomic, transcriptomic, etc., has specific processing methods to yield useable data points. It is important to understand what these data points represent. Are they values in an experimental sample relative to a control sample? Are they ratios of values in treated samples to untreated samples? Before any statistical analysis is performed, the researcher needs to know exactly what the data is representing so that the appropriate hypothesis can be tested.

4.3.2 Statistical Analysis Methods

The statistical analysis methods chosen to test hypotheses depend primarily on the type of data being analyzed, the definition of the dependent variable, and the sample size. Every statistical test has underlying assumptions that must be met for the method to be appropriate and there are specific tests developed for the analysis of small sample sizes. This section will describe the most commonly used statistical tests and define the appropriate use of them.

4.3.2.1 Data Type

The first characteristic that must be defined is the type of data one is applying the statistical test to. Typically, data falls into two broad categories, continuous and categorical. These names are descriptive of the data. A continuous variable is quantitative and measured on a continuous scale and can take on any value limited only by the measurement precision. The value of a continuous variable indicates some sort of amount. For example, protein expression reported in RFUs is a continuous variable that can take on any value between a minimum and a maximum, can be measured to the precision possible by the instrument, and indicates how much of the protein is expressed. Categorical data is, on the other hand, data that describes inclusion into a category. For example, experiment batch is an example of categorical data where the data defines which category or group the data point falls into, either it is a data point that was collected in batch A or it is a data point that was collected in batch B. The value of the category is not meaningful itself, i.e., batch A is not more or less of a batch than batch B. Categorical data can be further divided into nominal (categories that have no natural order) and ordinal (categories that have a natural order). This may seem a rather elementary description that most scientists will be familiar with, but the importance of knowing what type of data you have is sometimes forgotten. The statistical methods for the analysis of continuous data are completely different than those used for categorical data and they are, for the most part, not interchangeable. For a continuous variable, such as protein expression, it makes sense to compare mean levels between two groups. It does not make much sense, however, to compare mean levels of batch, a categorical variable, between two groups.

4.3.2.2 Dependent Versus Independent Variables

The distinction between the dependent variable and the independent variable(s) is an important element to a statistical analysis. While, mathematically, simple statistical tests between two variables, such as a Pearson correlation or a Chi-square test, do not differ based on which variable is considered dependent or independent, many statistical tests do differ. In addition, which variable is defined as the dependent variable dictates the interpretation of the statistical test.

Typically, an experiment has two classes of variables, the dependent variable that is tested and measured and the independent variables that are changed or controlled. The dependent variable is said to be "dependent" on the independent variable; as the independent variable changes, it exerts an effect on the dependent variable which is measured. We can say that the independent variable is the "input" to a statistical model which describes the change in the "output" or the dependent variable. This may seem an unnecessary complication for analysis, but this distinction defines how the results are interpreted. For an example of how interpretation is impacted, consider a simple experiment where one has a group of mice of varying ages and the interest is to evaluate if the expression of gene *ABC* changes as the mice get older. If expression of gene *ABC* is defined as the dependent variable and age as the independent variable, a correlation or linear regression would be used to assess the relationship. The conclusion, assuming a significant *p*-value, would indicate that age exerts an effect on gene expression; the gene expression level *depends* on how old the mouse is. If, on the other hand, age is defined as the dependent variable and expression as the independent variable, the same correlation or linear regression would be used. The conclusion from this analysis however, assuming a significant *p*-value, would indicate expression exerts an effect on age; the age of the mouse *depends* on what the expression of gene *ABC* is. These are two closely related conclusions, but they are saying something quite different. One is implying that the expression of gene *ABC* changes as the mice get older; the other is implying that age is dictated by the expression of gene *ABC*.

In more complicated statistical analyses where one has more than two variables, the definition of dependent/independent variables will mathematically change the statistical test. Here it is critically important to define the dependent variable appropriately so that the statistical test used is testing the hypothesis correctly. Before a choice of a statistical test is made, make sure that the dependent variable is defined correctly. Multiple regression models, where one has a single dependent variable and two or more independent variables, treat the dependent and independent variables differently. The model estimates calculated as part of a multiple regression model take into account all independent variables in the model and calculate their cumulative effect on the dependent variable.

4.3.2.3 Parametric Versus Nonparametric

The commonly used statistical tests described here fall into two groups, those that are parametric and those that are nonparametric. There are semi-parametric models that do not fit this paradigm; however, they are beyond the scope of this discussion. Parametric tests are those that make assumptions about the population distribution from which the sample is drawn; nonparametric tests do not make this assumption. Parametric tests use the data points as measured or after a systematic transformation has been applied, whereas nonparametric tests use the rank order of the data points (which value is the smallest, which is the next smallest, and so on) and ignore the distance between the points. Each parametric statistical test requires that the distribution assumption is met. If, and only if, one can show their data is drawn from the assumed population, the statistical test is valid.

When performing a statistical test that relies on an underlying distribution, such as the common student's *t*-test, the test is only valid if your sample has been drawn from a normal distribution. Applying a *t*-test to a highly skewed dependent variable violates the assumption of the test which can lead to either a reasonable conclusion or a biased conclusion. However, the difficulty is one doesn't know which. Before applying a statistical test, look at the dependent variable graphically and determine if it meets the assumptions. If it does not, a data transformation (i.e., log, square root, reciprocal transformation) can be applied, or a nonparametric test can be used.

4.3.2.4 Common Statistical Tests

When setting out to perform a statistical analysis, answer the three following questions of your data. First, what is the dependent variable and what is/are the independent variable(s)? Second, what type of data are the dependent and independent variables? Third, can we show that the dependent variable is drawn from a specific distribution, or do we not want to make that assumption? Once these three questions are answered, the choice of the appropriate statistical test is straightforward.

The table below is intended as a guide to the most commonly used statistical tests in omic experiments and those tests likely to be available using nonstatistical software. It is not an exhaustive list and analyses that require more complicated methods or designs should be performed with the guidance of a statistician.

Underlying distribution of dependent variable	Dependent variable	Independent variable	Test	Use
Normal	Continuous	Continuous	Pearson correlation	Assess the linear relationship between two continuous variables
Normal	Continuous	Categorical with two levels	Independent t-test for equal variances	Compare means between two independent groups; variances equal in both groups
Normal	Continuous	Categorical with two levels	Welch t-test for unequal variances	Compare means between two independent groups; variances unequal
Normal	Continuous	Categorical with three or more levels	ANOVA	Compare means between three or more independent groups
Normal	Continuous	Paired categorical with two levels	Paired t-test	Compare means between two paired groups (often pre/post)
Normal	Continuous	Paired categorical with three or more levels	Repeated measures ANOVA	Compare three or more means between groups (often pre/post)
Normal	Continuous	Any type/any number	Linear regression	Assess the effect the independent variable(s) exert on the dependent variable; can predict what the dependent variable would be if one knows the value for the independent variable(s)
Binomial	Categorical (two levels)	Any type/any number	Logistic regression	Assess the effect the independent variable(s) exert on the dependent variable; can predict what the dependent variable would be if one knows the value for the independent variable(s)
Binomial	Categorical (three or more levels)	Any type/any number	Ordinal or nominal regression	Assess the effect the independent variable(s) exert on the dependent variable; can predict what the dependent variable would be if one knows the value for the independent variable(s)
Chi2	Categorical	Categorical	Chi square test	Tests for an association between two categorical variables

(continued)

Underlying distribution of dependent variable	Dependent variable	Independent variable	Test	Use
Chi2	Categorical	Categorical	Fisher's exact test	Tests for an association between two categorical variables; used for small sample sizes
Chi2	Categorical	Categorical (2 levels)—Paired	McNemar's test	Tests for an association between two paired categorical variables
Chi2	Categorical	Categorical (3+ levels)—paired	Cochran's Q test	Tests for an association between two paired categorical variables where one variable has three or more levels
Binomial	Single proportion	Known proportion	Binomial test	Test deviations from a theoretically expected distribution of observations into two categories
Normal	Continuous	Continuous	Kolmogorov-Smirnov one sample test	Compares the distribution of the experimental variable to a normal distribution (normality test)
Normal	Continuous	Continuous	Kolmogorov-Smirnov two-sample test	Compares two continuous variables to determine if they arise from similar distributions
N/A	Continuous	Continuous	Spearman correlation	Assesses the linear relationship between two ranked continuous variables
N/A	Continuous	Categorical with two levels	Wilcoxon rank sum	Compare ranks between two independent groups
N/A	Continuous	Categorical with three or more levels	Kruskal-Wallis	Compare ranks between three or more independent groups
N/A	Continuous	Paired categorical with two levels	Wilcoxon sign rank	Compare ranks between two paired groups (often pre/post)
N/A	Continuous	Paired categorical with three or more levels	Freidman test	Compare three or more ranks between groups (often pre/post)

I highly recommend two reference books to guide the researcher in choosing the correct statistical test to use. These two references, [146, 147], discuss statistical tests in easy to understand language devoid of the heavy use of jargon. Pett discusses the choice of statistical methods specifically for small sample sizes, a common problem in omic experiments.

4.3.2.5 Statistical Model Complexity

Statistical models range from simple, where one is testing the relationship between two variables, to extremely complex where one can have many comparison groups, many time points, nested designs, and many covariates. As described above, it is important to make sure the sample size one has is adequate to test the hypothesis of interest. It is also important to make the best use of the number of samples one has, often a small number.

Consider one wants to look at expression levels between two groups of mice, treated and untreated, and also want to see if there is a sex difference. Analyze the data using a linear regression with expression as your dependent variable and with two independent variables, sex and treatment. This will allow you test the simultaneous effect of sex and treatment, and it will allow one, through the use of a sex $\times$ treatment interaction term, to determine if expression differences between treated and untreated are different in males and females. This is a much more efficient analysis than performing a comparison in males and females separately. It allows one to perform a direct comparison between males and females.

However, a more complex statistical model can require a larger sample size to adequately test the hypothesis. If one has many treatment groups or many doses of one treatment and are interested in evaluating the change in expression over many time points, a small sample size of six mice is probably not adequate. Here we must balance the need for using our data in the most efficient manner (using a more complex model) with the need for a model that can be adequately supported by our sample size.

4.3.2.6 Multiple Testing

The issue of multiple testing and the need to account for it is not a new phenomenon; however, it has gained a new prominence with the rise of high-throughput experiments where thousands, and sometimes millions, of statistical tests are performed on the same samples. Whereas before we concerned ourselves with the testing of 10 or 20 outcomes, now we are concerned with a number of statistical tests that is many orders of magnitude larger.

Each statistical test produces a *p*-value which most researchers are familiar with. Using the conventional significance level of 0.05 as our decision point, a *p*-value ≤ 0.05 will lead the researcher to reject the null hypothesis and conclude there is a significant effect; a *p*-value >0.05 will lead the researcher to accept the null hypothesis and conclude there is no evidence of a significant effect. In layman's terms, the *p*-value defines how likely the conclusion to accept or reject the null hypothesis has occurred by chance. A *p*-value of 0.05 indicates that there is a 1 in 20 chance that the conclusion occurred by chance (false positive) and a 19 in 20 chance that the conclusion is true. The smaller the *p*-value is, the less likely it occurred by chance. This concept applies to each statistical test performed.

Consider that for a single significance test performed at the conventional 0.05 level, we have a 5% chance (or a 1 in 20 chance) that our conclusion to reject the null hypothesis occurred by chance alone (is a false positive). Because of a mathematical property of probabilities, the likelihood of this conclusion occurring by chance (a false positive) increases with the number of tests performed. So that if one performs 20 statistical tests, there is a 64% chance of at least one conclusion being a false positive. This 64% of having at least one false positive is obviously much greater than 5%. Now consider when we perform thousands of statistical tests; the likelihood that at least one of our conclusions is false approaches 100%. Unfortunately, we cannot differentiate which of our significant conclusions are in fact false positives and which are true findings.

To combat this problem, much effort has been put into methods to control the error in an analysis [148] and these methods consider two different ways to solve this problem. The first methods control the family-wise error rate (FWER), the probability of finding one or more false positives among all of the statistical tests performed. The most common method, the Bonferroni method, adjusts the overall error rate to control the probability of at least one false positive *overall* rather than for each individual statistical test. Most researchers are probably familiar with the Bonferroni correction where the p-value used as our cutoff for accepting or rejecting the null hypothesis is adjusted by the number of hypotheses tested. For example, if the cutoff (i.e., significance level) for a single statistical test is 0.05, then the new cutoff for 10 statistical tests is 0.05/10 or 0.005. Now, only p-values that are ≤ 0.005 would be evidence to reject the null hypothesis and conclude there is a significant finding. Unfortunately, the Bonferroni correction specifically and the adjustment of the FWER generally are usually considered too conservative for most omic experiments [149] for reasons that go beyond the scope of this chapter. A second methodology, control of the false discovery rate (FDR), has been developed which controls the expected proportion of false positive among all of the significant findings (i.e., where the null hypothesis has been rejected). Many different methods have been developed to control each of these error rates (FWER and FDR) and they have been refined to resolve specific limitations. A review of the commonly used methods for both control of the FWER and the FDR can be found in Chen [150].

In any omic experiment where we are performing multiple statistical tests, as in testing the expression of thousands of genes, we need to account for multiple testing. If we do not, we cannot define which, if any, of our genes yielding a significant p-value are true findings and which are false positives.

4.3.2.7 Limitations and Future Directions

This chapter is intended as a resource to assist in the design and analysis of omic experiments; however, it is not a comprehensive guide. It introduces a high-level view of the bioinformatic techniques commonly used and an overview of the statistical methods typically used for omic experiments, but it does not discuss in detail the underlying mathematics or probability theory. It is advised that one consults with a statistician early in the experimental process, preferably during study design, to ensure that one is able to test the hypothesis of interest with the data collected. In addition, the guidance of a statistician is highly recommended for complex study designs as the methods used for their analysis go beyond the common methods discussed here. Some additional resources are given as references [151–161]. These resources describe statistical tests and their interpretation from a non-statistician perspective and include discussions of statistical analyses specifically for omic experiments.

Understanding the purpose of your experiment will determine what type of statistics are best suited and the inferential statistics described in this chapter may not even be needed. The purpose of inferential statistics is to make broad conclusions about a population from a smaller sample through the calculation of a p-value. If your purpose differs, maybe to describe a sample rather than make inferences about the broader population, the calculation of a p-value is unwarranted and unnecessary. If you have only technical replicates, all originating from the same biological sample, inferential statistics are unnecessary. Here any conclusions can only inform you about the population which consists of a single biological sample.

The statistical analysis of omic experiments is an evolving field where consensus on the most appropriate methods can be hard to find. As research continues, improved methods for the adjustment of multiple testing better suited to related outcomes will be developed and better methods of data integration from various sources and platforms will be defined. However, there remain several challenges in the analysis of omic data, whether it be integrating molecular and clinical data or adequately testing hypotheses so that the resulting conclusions are relevant for the population.

References

1. Hood, L., & Galas, D. (2003). The digital code of DNA. *Nature, 421*(6921), 444–448.
2. Dahm, R. (2008). Discovering DNA: Friedrich Miescher and the early years of nucleic acid research. *Human Genetics, 122*(6), 565–581.
3. Levy, S. E., & Myers, R. M. (2016). Advancements in next-generation sequencing. *Annual Review of Genomics and Human Genetics, 17*(1), 95–115.
4. Reis-Filho, J. S. (2009). Next-generation sequencing. *Breast Cancer Research, 11*(S3), S12.
5. Cock, P. J. A., Fields, C. J., Goto, N., Heuer, M. L., & Rice, P. M. (2010). The sanger FASTQ file format for sequences with quality scores, and the Solexa/Illumina FASTQ variants. *Nucleic Acids Research, 38*(6), 1767–1771.
6. Ewing, B., Hillier, L., Wendl, M. C., & Green, P. (1998). Base-calling of automated sequencer traces using phred. I. Accuracy assessment. *Genome Research, 8*(3), 175–185.
7. Ewing, B., & Green, P. (1998). Base-calling of automated sequencer traces using phred. II. Error probabilities. *Genome Research, 8*(3), 186–194.
8. Andrews, S. (2010). *FastQC a quality control tool for high throughput sequence data*. Retrieved November 25, 2018 from http://www.bioinformatics.babraham.ac.uk/projects/fastqc/.
9. Martin, M. (2011). Cutadapt removes adapter sequences from high-throughput sequencing reads. *EMBnet Journal, 17*(1), 10.

10. Joshi, N. A., & Fass, J. N. (2011). *Sickle: A sliding-window, adaptive, quality-based trimming tool for FastQ files*.
11. Li, H., & Durbin, R. (2009). Fast and accurate short read alignment with Burrows-Wheeler transform. *Bioinformatics, 25*(14), 1754–1760.
12. Adjeroh, D., Bell, T., & Mukherjee, A. (2008). *The Burrows-Wheeler transform: Data compression, suffix arrays, and pattern matching*. New York: Springer.
13. Lam, T. W., Sung, W. K., Tam, S. L., Wong, C. K., & Yiu, S. M. (2008). Compressed indexing and local alignment of DNA. *Bioinformatics, 24*(6), 791–797.
14. Li, H., et al. (2009). The sequence alignment/map format and SAMtools. *Bioinformatics, 25* (16), 2078–2079.
15. McKenna, A., et al. (2010). The genome analysis toolkit: A MapReduce framework for analyzing next-generation DNA sequencing data. *Genome Research, 20*(9), 1297–1303.
16. Garrison, E., & Marth, G. (2016). *Haplotype-based variant detection from short-read sequencing*.
17. Kobayashi, M., et al. (2017). Heap: A highly sensitive and accurate SNP detection tool for low-coverage high-throughput sequencing data. *DNA Research, 24*(4), 397–405.
18. Tattini, L., D'Aurizio, R., & Magi, A. (2015). Detection of genomic structural variants from next-generation sequencing data. *Frontiers in Bioengineering and Biotechnology, 3*, 92.
19. Chen, K., et al. (2009). BreakDancer: An algorithm for high-resolution mapping of genomic structural variation. *Nature Methods, 6*(9), 677–681.
20. Korbel, J. O., et al. (2009). PEMer: A computational framework with simulation-based error models for inferring genomic structural variants from massive paired-end sequencing data. *Genome Biology, 10*(2), R23.
21. Lee, S., Hormozdiari, F., Alkan, C., & Brudno, M. (2009). MoDIL: Detecting small indels from clone-end sequencing with mixtures of distributions. *Nature Methods, 6*(7), 473–474.
22. Magi, A., Tattini, L., Pippucci, T., Torricelli, F., & Benelli, M. (2012). Read count approach for DNA copy number variants detection. *Bioinformatics, 28*(4), 470–478.
23. Magi, A., et al. (2013). EXCAVATOR: Detecting copy number variants from whole-exome sequencing data. *Genome Biology, 14*(10), R120.
24. Abyzov, A., Urban, A. E., Snyder, M., & Gerstein, M. (2011). CNVnator: An approach to discover, genotype, and characterize typical and atypical CNVs from family and population genome sequencing. *Genome Research, 21*(6), 974–984.
25. Schröder, J., et al. (2014). Socrates: Identification of genomic rearrangements in tumour genomes by re-aligning soft clipped reads. *Bioinformatics, 30*(8), 1064–1072.
26. Karakoc, E., et al. (2012). Detection of structural variants and indels within exome data. *Nature Methods, 9*(2), 176–178.
27. Earl, D., et al. (2011). Assemblathon 1: A competitive assessment of de novo short read assembly methods. *Genome Research, 21*(12), 2224–2241.
28. Iqbal, Z., Caccamo, M., Turner, I., Flicek, P., & McVean, G. (2012). De novo assembly and genotyping of variants using colored de Bruijn graphs. *Nature Genetics, 44*(2), 226–232.
29. Nijkamp, J. F., van den Broek, M. A., Geertman, J.-M. A., Reinders, M. J. T., Daran, J.-M. G., & de Ridder, D. (2012). De novo detection of copy number variation by co-assembly. *Bioinformatics, 28*(24), 3195–3202.
30. Rausch, T., Zichner, T., Schlattl, A., Stutz, A. M., Benes, V., & Korbel, J. O. (2012). DELLY: Structural variant discovery by integrated paired-end and split-read analysis. *Bioinformatics, 28*(18), i333–i339.
31. Layer, R. M., Chiang, C., Quinlan, A. R., & Hall, I. M. (2014). LUMPY: a probabilistic framework for structural variant discovery. *Genome Biology, 15*(6), R84.
32. Wong, K., Keane, T. M., Stalker, J., & Adams, D. J. (2010). Enhanced structural variant and breakpoint detection using SVMerge by integration of multiple detection methods and local assembly. *Genome Biology, 11*(12), R128.
33. Jeffares, D. C., et al. (2017). Transient structural variations have strong effects on quantitative traits and reproductive isolation in fission yeast. *Nature Communications, 8*, 14061.

34. English, A. C., et al. (2015). Assessing structural variation in a personal genome—towards a human reference diploid genome. *BMC Genomics, 16*(1), 286.
35. Wang, K., Li, M., & Hakonarson, H. (2010). ANNOVAR: Functional annotation of genetic variants from high-throughput sequencing data. *Nucleic Acids Research, 38*(16), e164.
36. Sherry, S. T., et al. (2001). dbSNP: The NCBI database of genetic variation. *Nucleic Acids Research, 29*(1), 308–311.
37. MacDonald, J. R., Ziman, R., Yuen, R. K. C., Feuk, L., & Scherer, S. W. (2014). The database of genomic variants: A curated collection of structural variation in the human genome. *Nucleic Acids Research, 42*(Database issue), D986–D992.
38. Landrum, M. J., et al. (2018). ClinVar: Improving access to variant interpretations and supporting evidence. *Nucleic Acids Research, 46*(D1), D1062–D1067.
39. Rentzsch, P., Witten, D., Cooper, G. M., Shendure, J., & Kircher, M. (2018). CADD: Predicting the deleteriousness of variants throughout the human genome. *Nucleic Acids Research, 47*(D1), D886–D894.
40. Kircher, M., Witten, D. M., Jain, P., O'Roak, B. J., Cooper, G. M., & Shendure, J. (2014). A general framework for estimating the relative pathogenicity of human genetic variants. *Nature Genetics, 46*(3), 310–315.
41. Cingolani, P., et al. (2012). A program for annotating and predicting the effects of single nucleotide polymorphisms, SnpEff. *Fly, 6*(2), 80–92.
42. Cingolani, P., et al. (2012). Using *Drosophila melanogaster* as a model for genotoxic chemical mutational studies with a new program, SnpSift. *Frontiers in Genetics, 3*, 35.
43. Geoffroy, V., et al. (2018). AnnotSV: An integrated tool for structural variations annotation. *Bioinformatics, 34*(20), 3572–3574.
44. Freeman, W. M., Walker, S. J., & Vrana, K. E. (1999). Quantitative RT-PCR: Pitfalls and potential. *BioTechniques, 26*(1), 112–125.
45. Bumgarner, R. (2013). Overview of DNA microarrays: Types, applications, and their future. *Current Protocols in Molecular Biology, 101*(1), 22–21.
46. Solomon, M. J., Larsen, P. L., & Varshavsky, A. (1988). Mapping protein-DNA interactions in vivo with formaldehyde: Evidence that histone H4 is retained on a highly transcribed gene. *Cell, 53*(6), 937–947.
47. Van Gelder, R. N., von Zastrow, M. E., Yool, A., Dement, W. C., Barchas, J. D., & Eberwine, J. H. (1990). Amplified RNA synthesized from limited quantities of heterogeneous cDNA. *Proceedings of the National Academy of Sciences of the United States of America, 87*(5), 1663–1667.
48. Shalon, D., Smith, S. J., & Brown, P. O. (1996). A DNA microarray system for analyzing complex DNA samples using two-color fluorescent probe hybridization. *Genome Research, 6* (7), 639–645.
49. Ritchie, M. E., et al. (2015). Limma powers differential expression analyses for RNA-sequencing and microarray studies. *Nucleic Acids Research, 43*(7), e47.
50. Gautier, L., Cope, L., Bolstad, B. M., & Irizarry, R. A. (2004). Affy--analysis of Affymetrix GeneChip data at the probe level. *Bioinformatics, 20*(3), 307–315.
51. Dunning, M. J., Smith, M. L., Ritchie, M. E., & Tavare, S. (2007). Beadarray: R classes and methods for Illumina bead-based data. *Bioinformatics, 23*(16), 2183–2184.
52. Bolstad, B. M., Irizarry, R., Astrand, M., & Speed, T. P. (2003). A comparison of normalization methods for high density oligonucleotide array data based on variance and bias. *Bioinformatics, 19*(2), 185–193.
53. Carvalho, B. S., & Irizarry, R. A. (2010). A framework for oligonucleotide microarray preprocessing. *Bioinformatics, 26*(19), 2363–2367.
54. Warnes, G. R., Bolker, B., Bonebakker, L., Gentleman, R., Huber, W., & Liaw, A. (2009). gplots: Various R programming tools for plotting data. *R Packag. version 2*.
55. Student. (1908). The probable error of a mean. *Biometrika*. Retreived May 07, 2016, from http://seismo.berkeley.edu/~kirchner/eps_120/Odds_n_ends/Students_original_paper.pdf.

56. Fisher, R. A. (1919). XV.—The correlation between relatives on the supposition of Mendelian inheritance. *Transactions of the Royal Society of Edinburgh, 52*(02), 399–433.
57. Benjamini, Y., & Hochberg, Y. (1995). Controlling the false discovery rate: A practical and powerful approach to multiple testing. *Journal of the Royal Statistical Society, Series B, 57*(1), 289–300.
58. Bonferroni, C. (1936). Teoria statistica delle classi e calcolo delle probabilita. *Pubblicazioni del R Istituto Superiore di Scienze Economiche e Commericiali di Firenze, 8*, 3–62.
59. Schadt, E. E., Turner, S., & Kasarskis, A. (2010). A window into third-generation sequencing. *Human Molecular Genetics, 19*(R2), R227–R240.
60. Mikheyev, A. S., & Tin, M. M. Y. (2014). A first look at the Oxford Nanopore MinION sequencer. *Molecular Ecology Resources, 14*(6), 1097–1102.
61. Eisenstein, M. (2012). Oxford Nanopore announcement sets sequencing sector abuzz. *Nature Biotechnology, 30*(4), 295–296.
62. Kim, D., Pertea, G., Trapnell, C., Pimentel, H., Kelley, R., & Salzberg, S. L. (2013). TopHat2: Accurate alignment of transcriptomes in the presence of insertions, deletions and gene fusions. *Genome Biology, 14*(4), R36.
63. Trapnell, C., et al. (2012). Differential gene and transcript expression analysis of RNA-seq experiments with TopHat and Cufflinks. *Nature Protocols, 7*(3), 562–578.
64. Trapnell, C., et al. (2010). Transcript assembly and quantification by RNA-Seq reveals unannotated transcripts and isoform switching during cell differentiation. *Nature Biotechnology, 28*(5), 511–515.
65. Langmead, B., & Salzberg, S. L. (2012). Fast gapped-read alignment with Bowtie 2. *Nature Methods, 9*(4), 357–359.
66. Ferragina, P., & Manzini, G. (2001). An experimental study of a compressed index. *Information Sciences, 135*(1–2), 13–28.
67. Dobin, A., et al. (2013). STAR: Ultrafast universal RNA-seq aligner. *Bioinformatics, 29*(1), 15–21.
68. Kim, D., Langmead, B., & Salzberg, S. L. (2015). HISAT: A fast spliced aligner with low memory requirements. *Nature Methods, 12*(4), 357–360.
69. Pertea, M., Kim, D., Pertea, G. M., Leek, J. T., & Salzberg, S. L. (2016). Transcript-level expression analysis of RNA-seq experiments with HISAT, StringTie and Ballgown. *Nature Protocols, 11*(9), 1650–1667.
70. Pertea, M., Pertea, G. M., Antonescu, C. M., Chang, T.-C., Mendell, J. T., & Salzberg, S. L. (2015). StringTie enables improved reconstruction of a transcriptome from RNA-seq reads. *Nature Biotechnology, 33*(3), 290–295.
71. Frazee, A. C., Pertea, G., Jaffe, A. E., Langmead, B., Salzberg, S. L., & Leek, J. T. (2015). Ballgown bridges the gap between transcriptome assembly and expression analysis. *Nature Biotechnology, 33*(3), 243–246.
72. Wang, L., Wang, S., & Li, W. (2012). RSeQC: Quality control of RNA-seq experiments. *Bioinformatics, 28*(16), 2184–2185.
73. Mortazavi, A., Williams, B. A., McCue, K., Schaeffer, L., & Wold, B. (2008). Mapping and quantifying mammalian transcriptomes by RNA-Seq. *Nature Methods, 5*(7), 621–628.
74. Li, B., Ruotti, V., Stewart, R. M., Thomson, J. A., & Dewey, C. N. (2010). RNA-Seq gene expression estimation with read mapping uncertainty. *Bioinformatics, 26*(4), 493–500.
75. Li, B., & Dewey, C. N. (2011). RSEM: Accurate transcript quantification from RNA-Seq data with or without a reference genome. *BMC Bioinformatics, 12*(1), 323.
76. Dempster, A. P., Laird, N. M., & Rubin, D. B. (1976). Maximum likelihood from incomplete data via the EM algorithm. *Journal of the Royal Statistical Society: Series B (Methodological), 39*(1), 1–22.
77. Anders, S., Pyl, P. T., & Huber, W. (2015). HTSeq--a Python framework to work with high-throughput sequencing data. *Bioinformatics, 31*(2), 166–169.
78. Liao, Y., Smyth, G. K., & Shi, W. (2014). featureCounts: An efficient general purpose program for assigning sequence reads to genomic features. *Bioinformatics, 30*(7), 923–930.

79. Lawrence, M., et al. (2013). Software for computing and annotating genomic ranges. *PLoS Computational Biology, 9*(8), e1003118.
80. Soneson, C., Love, M. I., & Robinson, M. D. (2015). Differential analyses for RNA-seq: Transcript-level estimates improve gene-level inferences. *F1000Research, 4*, 1521.
81. Robinson, M. D., McCarthy, D. J., & Smyth, G. K. (2010). edgeR: A bioconductor package for differential expression analysis of digital gene expression data. *Bioinformatics, 26*(1), 139–140.
82. Love, M. I., Huber, W., & Anders, S. (2014). Moderated estimation of fold change and dispersion for RNA-seq data with DESeq2. *Genome Biology, 15*(12), 550.
83. Nelder, J. A., & Wedderburn, R. W. M. (1972). Generalized linear models. *Journal of the Royal Statistical Society, Series A, 135*(3), 370.
84. Wald, A. (1945). Sequential tests of statistical hypotheses. *Annals of Mathematical Statistics, 16*(2), 117–186.
85. Feng, J., Meyer, C. A., Wang, Q., Liu, J. S., Shirley Liu, X., & Zhang, Y. (2012). GFOLD: A generalized fold change for ranking differentially expressed genes from RNA-seq data. *Bioinformatics, 28*(21), 2782–2788.
86. Tarazona, S., et al. (2015). Data quality aware analysis of differential expression in RNA-seq with NOISeq R/Bioc package. *Nucleic Acids Research, 43*(21), e140.
87. Toedling, J., & Huber, W. (2008). Analyzing ChIP-chip data using bioconductor. *PLoS Computational Biology, 4*(11), e1000227.
88. Toedling, J., Sklyar, O., & Huber, W. (2007). Ringo – an R/bioconductor package for analyzing ChIP-chip readouts. *BMC Bioinformatics, 8*(1), 221.
89. Durinck, S., et al. (2005). BioMart and bioconductor: A powerful link between biological databases and microarray data analysis. *Bioinformatics, 21*(16), 3439–3440.
90. Alexa, A., Rahnenfuhrer, J., & Lengauer, T. (2006). Improved scoring of functional groups from gene expression data by decorrelating GO graph structure. *Bioinformatics, 22*(13), 1600–1607.
91. Zhang, Y., et al. (2008). Model-based Analysis of ChIP-Seq (MACS). *Genome Biology, 9*(9), R137.
92. Xu, S., Grullon, S., Ge, K., & Peng, W. (2014). Spatial clustering for identification of ChIP-enriched regions (SICER) to map regions of histone methylation patterns in embryonic stem cells. *Methods in Molecular Biology, 1150*, 97.
93. Hayatsu, H. (2008). Discovery of bisulfite-mediated cytosine conversion to uracil, the key reaction for DNA methylation analysis – a personal account. *Proceedings of the Japan Academy. Series B, Physical and Biological Sciences, 84*(8), 321–330.
94. Morris, T. J., et al. (2014). ChAMP: 450k chip analysis methylation pipeline. *Bioinformatics, 30*(3), 428–430.
95. Tian, Y., et al. (2017). ChAMP: Updated methylation analysis pipeline for illumina BeadChips. *Bioinformatics, 33*(24), 3982–3984.
96. Aryee, M. J., et al. (2014). Minfi: A flexible and comprehensive bioconductor package for the analysis of infinium DNA methylation microarrays. *Bioinformatics, 30*(10), 1363–1369.
97. Leek, J. T., Johnson, W. E., Parker, H. S., Jaffe, A. E., & Storey, J. D. (2012). The sva package for removing batch effects and other unwanted variation in high-throughput experiments. *Bioinformatics, 28*(6), 882–883.
98. Carson Sievert, P. T. I., Parmer, C., Hocking, T., Chamberlain, S., Ram, K., Corvellec, M., & Despouy, P. (2018). Create interactive web graphics via 'plotly.js' [R package plotly version 4.8.0]. *Comprehensive R Archive Network (CRAN).*
99. Krueger, F., & Andrews, S. R. (2011). Bismark: A flexible aligner and methylation caller for bisulfite-seq applications. *Bioinformatics, 27*(11), 1571–1572.
100. Chen, P.-Y., Cokus, S. J., & Pellegrini, M. (2010). BS Seeker: Precise mapping for bisulfite sequencing. *BMC Bioinformatics, 11*(1), 203.

101. Kreck, B., Marnellos, G., Richter, J., Krueger, F., Siebert, R., & Franke, A. (2012). B-SOLANA: An approach for the analysis of two-base encoding bisulfite sequencing data. *Bioinformatics, 28*(3), 428–429.
102. Frith, M. C., Mori, R., & Asai, K. (2012). A mostly traditional approach improves alignment of bisulfite-converted DNA. *Nucleic Acids Research, 40*(13), e100.
103. Saito, Y., Tsuji, J., & Mituyama, T. (2014). Bisulfighter: Accurate detection of methylated cytosines and differentially methylated regions. *Nucleic Acids Research, 42*(6), e45.
104. Xi, Y., & Li, W. (2009). BSMAP: Whole genome bisulfite sequence MAPping program. *BMC Bioinformatics, 10*(1), 232.
105. Assenov, Y., Müller, F., Lutsik, P., Walter, J., Lengauer, T., & Bock, C. (2014). Comprehensive analysis of DNA methylation data with RnBeads. *Nature Methods, 11*(11), 1138–1140.
106. Saito, Y., & Mituyama, T. (2015). Detection of differentially methylated regions from bisulfite-seq data by hidden Markov models incorporating genome-wide methylation level distributions. *BMC Genomics, 16*(Suppl 12), S3.
107. Song, Q., Decato, B., Hong, E. E., Zhou, M., & Fang, F. (2013). A reference methylome database and analysis pipeline to facilitate integrative and comparative epigenomics. *PLoS One, 8*(12), 81148.
108. Hansen, K. D., Langmead, B., & Irizarry, R. A. (2012). BSmooth: From whole genome bisulfite sequencing reads to differentially methylated regions. *Genome Biology, 13*(10), R83.
109. Hebestreit, K., Dugas, M., & Klein, H.-U. (2013). Detection of significantly differentially methylated regions in targeted bisulfite sequencing data. *Bioinformatics, 29*(13), 1647–1653.
110. Wreczycka, K., Gosdschan, A., Yusuf, D., Grüning, B., Assenov, Y., & Akalin, A. (2017). Strategies for analyzing bisulfite sequencing data. *Journal of Biotechnology, 261*, 105–115.
111. Tsuji, J., & Weng, Z. (2015). Evaluation of preprocessing, mapping and postprocessing algorithms for analyzing whole genome bisulfite sequencing data. *Briefings in Bioinformatics, 17*(6), bbv103.
112. Eberwine, J., et al. (1992). Analysis of gene expression in single live neurons. *Proceedings of the National Academy of Sciences of the United States of America, 89*(7), 3010–3014.
113. Hwang, B., Lee, J. H., & Bang, D. (2018). Single-cell RNA sequencing technologies and bioinformatics pipelines. *Experimental and Molecular Medicine, 50*(8), 96.
114. Van Der Maaten, L., & Hinton, G. (2008). Visualizing data using t-SNE. *Journal of Machine Learning Research, 9*, 2579–2605.
115. Butler, A., Hoffman, P., Smibert, P., Papalexi, E., & Satija, R. (2018). Integrating single-cell transcriptomic data across different conditions, technologies, and species. *Nature Biotechnology, 36*(5), 411–420.
116. Afgan, E., et al. (2018). The Galaxy platform for accessible, reproducible and collaborative biomedical analyses: 2018 update. *Nucleic Acids Research, 46*(W1), W537–W544.
117. Ashburner, M., et al. (2000). Gene ontology: Tool for the unification of biology. *Nature Genetics, 25*(1), 25–29.
118. Huang, D. W., Sherman, B. T., & Lempicki, R. A. (2009). Bioinformatics enrichment tools: Paths toward the comprehensive functional analysis of large gene lists. *Nucleic Acids Research, 37*(1), 1–13.
119. Fisher, R. A. (1922). On the interpretation of χ^2 from contingency tables, and the calculation of P. *Journal of the Royal Statistical Society, 85*(1), 87.
120. Ludbrook, J. (2008). Analysis of 2×2 tables of frequencies: Matching test to experimental design. *International Journal of Epidemiology, 37*(6), 1430–1435.
121. Huang, D. W., Sherman, B. T., & Lempicki, R. A. (2009). Systematic and integrative analysis of large gene lists using DAVID bioinformatics resources. *Nature Protocols, 4*(1), 44–57.
122. Falcon, S., & Gentleman, R. (2007). Using GOstats to test gene lists for GO term association. *Bioinformatics, 23*(2), 257–258.
123. Maere, S., Heymans, K., & Kuiper, M. (2005). BiNGO: A cytoscape plugin to assess overrepresentation of gene ontology categories in biological networks. *Bioinformatics, 21*(16), 3448–3449.

124. Subramanian, A., et al. (2005). Gene set enrichment analysis: A knowledge-based approach for interpreting genome-wide expression profiles. *Proceedings of the National Academy of Sciences, 102*(43), 15545–15550.
125. Lee, H. K., Braynen, W., Keshav, K., & Pavlidis, P. (2005). ErmineJ: Tool for functional analysis of gene expression data sets. *BMC Bioinformatics, 6*(1), 269.
126. Al-Shahrour, F., et al. (2007). From genes to functional classes in the study of biological systems. *BMC Bioinformatics, 8*, 114.
127. Nam, D., Kim, S.-B., Kim, S.-K., Yang, S., Kim, S.-Y., & Chu, I.-S. (2006). ADGO: Analysis of differentially expressed gene sets using composite GO annotation. *Bioinformatics, 22*(18), 2249–2253.
128. Nogales-Cadenas, R., et al. (2009). GeneCodis: Interpreting gene lists through enrichment analysis and integration of diverse biological information. *Nucleic Acids Research, 37*(Web Server issue), W317–W322.
129. Kanehisa, M., & Goto, S. (2000). KEGG: Kyoto encyclopedia of genes and genomes. *Nucleic Acids Research, 28*(1), 27–30.
130. Finn, R. D., et al. (2014). Pfam: The protein families database. *Nucleic Acids Research, 42* (Database issue), D222–D230.
131. Matys, V., et al. (2003). TRANSFAC: Transcriptional regulation, from patterns to profiles. *Nucleic Acids Research, 31*(1), 374–378.
132. Warde-Farley, D., et al. (2010). The GeneMANIA prediction server: Biological network integration for gene prioritization and predicting gene function. *Nucleic Acids Research, 38* (Web Server issue), W214–W220.
133. Stark, C., Breitkreutz, B.-J., Reguly, T., Boucher, L., Breitkreutz, A., & Tyers, M. (2006). BioGRID: A general repository for interaction datasets. *Nucleic Acids Research, 34*(Database issue), D535–D539.
134. Zhang, B., & Horvath, S. (2005). A general framework for weighted gene co-expression network analysis. *Statistical Applications in Genetics and Molecular Biology, 4*(1), Article17.
135. Langfelder, P., & Horvath, S. (2008). WGCNA: An R package for weighted correlation network analysis. *BMC Bioinformatics, 9*(1), 559.
136. Ewels, P., Magnusson, M., Lundin, S., & Käller, M. (2016). MultiQC: Summarize analysis results for multiple tools and samples in a single report. *Bioinformatics, 32*(19), 3047–3048.
137. Gregory, R., Warnes, R., Bolker, B., Bonebakker, L., Gentleman, M., Liaw, W. H. A., Lumley, T., Maechler, B., Magnusson, A., Moeller, S., Schwartz, M., & Venables, B. (2016). Various R programming tools for plotting data. *R Package Version, 2*(4), 1.
138. Walter, W., Sánchez-Cabo, F., & Ricote, M. (2015). GOplot: An R package for visually combining expression data with functional analysis. *Bioinformatics, 31*(17), 2912–2914.
139. Ghosh, D., & Poisson, L. M. (2009). "Omics" data and levels of evidence for biomarker discovery. *Genomics, 93*, 13–16.
140. Wheelock, A. M., & Wheelock, C. E. (2013). Trials and tribulations of 'omics data analysis: Assessing quality of SIMCA-based multivariate models using examples from pulmonary medicine. *Molecular BioSystems, 9*, 2589.
141. Kraus, L. (2015). Editorial: Would you like a hypothesis with those data? Omics and the age of discovery science. *Molecular Endocrinology, 29*(11), 1531–1534.
142. Vaux, D. L., Fidler, F., & Cumming, G. (2012). Replicates and repeats—What is the difference and is it significant? A brief discussion of statistics and experimental design. *EMBO Reports, 13*(4), 291.
143. Bell, G. (2016). Comment: Replicates and repeats. *BMC Biology, 14*, 28.
144. Whitley, E., & Ball, J. (2002). Statistics review 4: Sample size calculations. *Critical Care, 6* (4), 335.
145. Billoir, E., Navratil, V., & Blaise, B. J. (2015). Sample size calculation in metabolic phenotyping studies. *Briefings in Bioinformatics, 16*(5), 813–819.
146. Urdan, T. C. (2010). *Statistics in plain English* (3rd ed.). New York: Routledge.

147. Pett, M. A. (1997). *Nonparametric statistics for health care research: Statistics for small samples and unusual distributions*. Thousand Oaks, CA: Sage.
148. Storey, J. D. (2002). A direct approach to false discovery rates. *Journal of the Royal Statistical Society. Series B, 64*(Part 3), 479–498.
149. Feise, R. J. (2002). Do multiple outcome measures require p-value adjustment? *BMC Medical Research Methodology, 2*, 8.
150. Chen, S. Y., Feng, Z., & Yi, X. (2017). A general introduction to adjustment for multiple comparisons. *Journal of Thoracic Disease, 9*(6), 1725–1729.
151. Forshed, J. (2017). Experimental design in clinical 'omics biomarker discovery. *Journal of Proteome Research, 16*, 3954–3960.
152. Guyatt, G., Jaeschke, R., Heddle, N., Cook, D., Shannon, H., & Walter, S. (1995). Basic statistics for clinicians: 1. Hypothesis testing. *CMAJ, 152*(1), 27–32.
153. Guyatt, G., Jaeschke, R., Heddle, N., Cook, D., Shannon, H., & Walter, S. (1995). Basic statistics for clinicians: 2. Interpreting study results: Confidence intervals. *CMAJ, 152*(2), 169–173.
154. Guyatt, G., Walkter, S., Shannon, H., Cook, D., Jaeschke, R., & Heddle, N. (1995). Basic statistics for clinicians: 4. Correlation and regression. *CMAJ, 152*(4), 497–504.
155. Hanley, J. A., & Moodie, E. E. M. (2011). Sample size, precision and power calculations: A unified approach. *Journal of Biometrics and Biostatistics, 2*, 5.
156. Ioannidis, J. P. A., Tarone, R., & McLaughlin, J. K. (2011). The false-positive to false-negative ratio in epidemiologic studies. *Epidemiology, 22*(4), 450–456.
157. Jarschke, R., Guyatt, G., Shannon, H., Walter, S., Cook, D., & Heddle, N. (1995). Basic statistics for clinicians: 3. Assessing the effects of treatment: Measures of association. *CMAJ, 152*(3), 351–357.
158. Mazzocchi, F. (2015). Could big data be the end of theory in science? A few remarks on the epistemology of data-driven science. *EMBO Reports, 16*(10), 1250–1255.
159. Rajasundaram, D., & Selbig, J. (2016). More effort — More results: Recent advances in integrative 'omics' data analysis. *Current Opinion in Plant Biology, 30*, 57–61.
160. Senn, S., & Bretz, F. (2007). Power and sample size when multiple endpoints are considered. *Pharmaceutical Statistics, 6*, 161–170.
161. Signe, A., Esteban, F. J., Stavreus-Evers, A., Simon, C., Giudice, L., Lessey, B. A., Horcajadas, J. A., Macklon, N. S., D'Hooghe, T., Campoy, C., Fauser, B. C., Salamonsen, L. A., & Salumets, A. (2014). Guidelines for the design, analysis and interpretation of 'omics' data: Focus on human endometrium. *Human Reproduction Update, 20*(1), 12–28.

Part II
Transcriptomic

Chapter 5
Transcriptomic Approaches for Muscle Biology and Disorders

Poching Liu, Surajit Bhattacharya, and Yi-Wen Chen

The transcriptome refers to all RNA molecules transcribed from our genome, including messenger RNA (mRNA), ribosomal RNA (rRNA), transfer RNA (tRNA), as well as other regulatory noncoding RNAs. Some well-known regulatory noncoding RNA molecules are long noncoding RNA (lncRNA), microRNA (miRNA), small nuclear RNA (snRNA), and small nucleolar RNA (snoRNA). Among the different types of RNA, the protein-coding mRNA received the most attention, particularly when the tools for RNA profiling, or gene expression profiling, were first developed in the late twentieth century. Since then, different tools and sample preparation techniques have been developed to target various groups of RNAs for study. This chapter will focus on mRNA profiling approaches primarily; however, the same or modified technologies can be used to study other RNA groups. The approach for studying miRNA, miRNA profiling, is described in a separate chapter.

Transcriptomic studies are conducted to understand how transcriptome differences and changes contribute to biological functions and diseases. For skeletal muscle research, a large number of studies have been published in the past 20 years using arrays and sequencing approaches to investigate (1) basic muscle

P. Liu
DNA Sequencing and Genomics Core, National Heart, Lung and Blood Institute, National Institutes of Health, Bethesda, MD, USA

S. Bhattacharya
Center for Genetic Medicine Research, Children's National Medical Center, Washington, DC, USA

Y.-W. Chen (✉)
Center for Genetic Medicine Research, Children's National Health System, Children's Research Institute, Washington, DC, USA

Department of Genomics and Precision Medicine, School of Medicine and Health Science, George Washington University, Washington, DC, USA
e-mail: YChen@childrensnational.org

J. G. Burniston, Y.-W. Chen (eds.), *Omics Approaches to Understanding Muscle Biology*, Methods in Physiology, https://doi.org/10.1007/978-1-4939-9802-9_5

biology [1–5]; (2) molecular responses to physiological and environmental stimuli [6–10]; (3) effects of aging on muscles [11, 12]; (4) disease mechanism of muscle disorders [13–18]; (5) molecular changes in muscles of non-muscle diseases [19–21]; and (6) molecular responses to therapeutic interventions [22–24]. Most of these transcriptome data sets have been deposited into public databases, such as the Gene Expression Omnibus (GEO) database hosted by the National Center for Biotechnology Information (NCBI), National Institutes of Health (NIH) (https://www.ncbi.nlm.nih.gov/geo/). This provides a rich source of information for researchers today. For example, one can query and download data of interest and conduct further analyses. In this chapter, we will focus on the most commonly used platforms and approaches to generate these data sets.

5.1 Theory and History of the Technique

Gene expression is the process wherein information from a gene is used in the production of a functional gene product. When the final product is a protein, the process involves "transcription" which produces mRNA using DNA as the template and "translation" which produces proteins using the information provided by the mRNA transcripts. The size of human genome is estimated to be 3.3 billion base pairs, and approximately 5% of it can be transcribed and produce RNA products. Of the whole transcriptome, only ~4% of the transcribed RNAs are protein-coding mRNAs [25]. The amount of mRNA changes slightly depending on the cell types. Dividing cells that are more active transcriptionally produce more mRNA transcripts compared to terminally differentiated cells, such as myofibers. In addition to mRNA quantity, gene expression profiles are different in different cell types and cell states. Controlling which genes are expressed at a given time enables the cell to control its size, shape, and functions. In other words, while all cells carry the same DNA content, only a small portion of the DNA is actively transcribed at any one point in time. Among the transcripts produced, only a small portion is protein-coding RNAs. Transcriptomic approaches allow us to investigate differences in these transcripts in muscle tissues and cells in different physiological conditions and disease states. By comparing samples of interests to the controls, one may discover important molecular changes and pathways that play critical roles in the conditions or diseases under study. Because many regulatory steps in addition to transcription are involved in regulating protein synthesis and function, one should not make conclusions solely based on mRNA profiling data. Previous studies showed that mRNA expression levels positively correlate with protein levels; however, the change of mRNA level does not fully explain changes in proteins because posttranslational modifications such as phosphorylation and methylation may have further effects on protein function [26–30].

Gene expression profiling measures RNA transcripts at a given moment, which provides a "snapshot" of the gene activities in cells in a specific state or condition. In general, methods for determining RNA transcript levels can be based on (1) transcript visualization, (2) transcript hybridization, or (3) transcript sequencing. Because RNA molecules are prone to degradation, they are usually reverse-transcribed into cDNA (complementary DNA) before further processing. However, there are exceptions which will be discussed later. Polymerase chain reaction (PCR) arrays use fluorescent dyes to detect and quantify the amount of PCR amplicons of each transcript in a sample. Microarrays use cDNA or oligonucleotide probes to detect the transcripts by hybridization methods. The next-generation sequencing approach directly sequences the transcripts and determines the expression level of each transcript by the number of reads. Earlier technologies for expression profiling (cDNA arrays or PCR-based arrays) measure the expression of hundreds to thousands of genes at a time. These arrays can be made in-house or custom-made by companies. Afterward, microarrays were commercially developed by several companies to increase coverage, throughput, sensitivity, and specificity. Some of the platforms are capable of surveying tens of thousands of transcripts and allow genome-wide profiling. Commonly used platforms include Affymetrix GeneChip® Microarrays, Illumina High-Density Silica Bead-Based Microarrays, and Agilent Expression Microarrays. For these microarray platforms, oligonucleotide probes at various lengths accompanied by different array designs are used to capture the target transcripts by hybridization methods. Transcripts that have not been discovered or are not included on the microarrays will not be detected or measured using these platforms. This limitation was resolved by the next-generation sequencing technologies, which first became available at the beginning of the 2000s [31, 32]. Different from the traditional Sanger sequencing, next-generation sequencing (NGS) does not target specific sequences. Instead, all transcripts in a sample are sequenced. This unbiased approach in theory can survey the whole transcriptome including both coding and noncoding RNA. However, due to the limitation of how many reads can be obtained in one run and one may only be interested in a specific type of RNA transcripts, different protocols, including primer types and enrichment methods, were developed to examine specific types of RNA transcripts.

5.2 Major Applications

5.2.1 Hybridization-Based Platforms

The principle of microarrays is nucleic acid hybridization, in which two complementary strands of DNA or RNA molecules join to form a double-stranded molecule, following the complimentary base pairing rules: adenine (A) pairs with thymine (T) and cytosine (C) pairs with guanine (G). Northern blotting uses probes that are complementary to the target RNA to detect an RNA that is immobilized on a nitrocellulose or nylon membrane. To detect many RNA transcripts at the same time, instead of immobilizing the RNA samples, many different probes are immobilized, i.e., spotted onto the membrane or glass slide and then used to detect the target RNA transcripts in samples adding to the immobilized probes. In this case, RNA

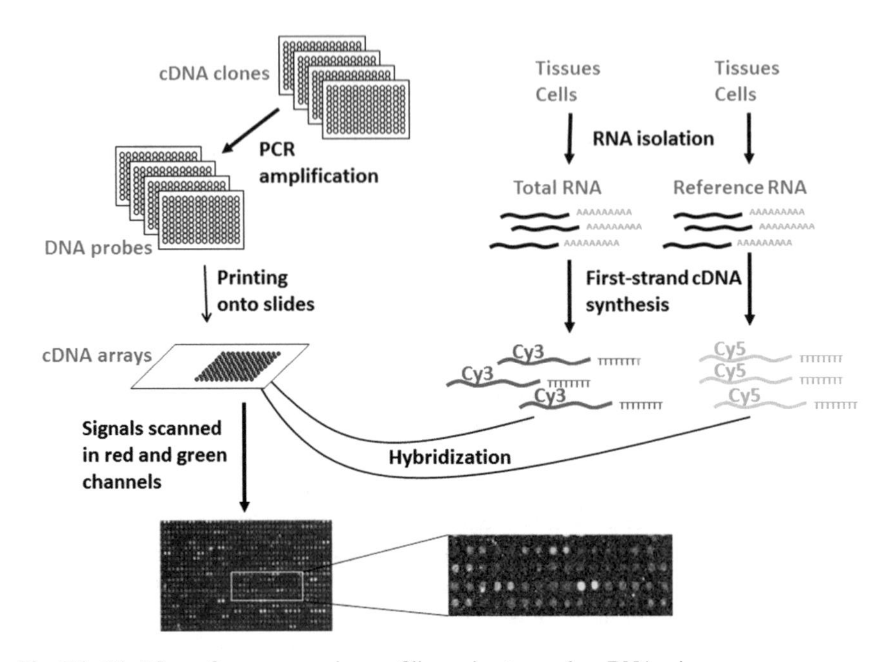

Fig. 5.1 Workflow of gene expression profiling using two-color cDNA microarrays

transcripts are reverse-transcribed into cDNA and then hybridized to the probes on the slides. The cDNA fragments are fluorescently labeled for visualization. The prototype of the cDNA microarray made by spotting known cDNAs into a 96-well microplate was reported in 1995, on which cDNAs were printed by robot [33]. Differential expression of 45 Arabidopsis genes were measured simultaneously using two-color fluorescence approach. In1996, 870 cDNAs were spotted on glass slides to determine gene expression for cancer classification [34]. Two different colors are used for the platform, the sample of interest is labeled with one color (e.g., Cy3), and a reference sample is labeled with a different color (e.g., Cy5). Both samples are hybridized to the same array and the ratio between the experiment and the reference samples are used for further data analyses. This design allows proper normalization and comparisons among different arrays (Fig. 5.1). One of the earliest expression profiling studies done in skeletal muscles was reported in 1996, in which membranes with cDNA clones spotted on them were used [5]. The early versions of cDNA arrays provide platforms for examining hundreds of transcripts simultaneously. There are several issues associated with the cDNA arrays, such as sensitivity, specificity, normalization, reproducibility, throughput, and coverage which were addressed in later generations of microarrays. Next, we describe three commercial platforms that are commonly used for expression profiling studies. A comparison of the three platforms is in Table 5.1 at the end of this section.

Table 5.1 Features of Affymetrix, Illumina, and Agilent microarrays

Manufacture	Affymetrix	Illumina	Agilent
Feature size (μM)	10	3	50
Oligonucleotide length	25 nt	50 nt	60 nt
Number of oligonucleotides/gene	10–20	1–2	1–2
Number of features/array	Up to 10,00,000	Up to 40,00,000	Up to 2,00,000
Custom flexibility	Limited	Moderate	Excellent

5.2.1.1 Affymetrix

Affymetrix makes GeneChip arrays for transcriptome analysis using millions of 25-base DNA probes that have been synthesized directly onto a glass chip using light-directed oligonucleotide synthesis method [35]. The probes are used to "probe" the sample for target RNA segments. Hybridization is the basis for the detections. Because the 25-mer is relatively short and can potentially bind to more than one transcript, instead of one probe, a set of 11 probes are designed for each transcript. The intensity data from all 11 probes are analyzed to determine whether the transcript is detected (called "present") or not detected (called "absent") and give a measure of the level of expression. In an experiment, the mRNA is reverse-transcribed into double-stranded cDNA. The cDNA is then used as the template for in vitro transcription. The RNA produced is known as cRNA, in which the uracil bases are labeled with biotin. The fragments of biotin-labeled cRNA are then loaded into the array so that the sample can hybridize with the probes on the glass chip of the array. Afterward the array is washed to remove RNA that has not hybridized to the probes. The hybridization is then visualized using streptavidin-linked fluorescent dyes that bind to the biotin-labeled cRNA. The array is scanned with a laser scanner and the image is analyzed to determine which transcripts are detected and how much of each transcript is present.

In 2000, Affymetrix GeneChip® HuGeneFL Arrays were used to identify transcriptomic changes underlying disease mechanisms of muscular dystrophies [18]. In this study, approximately 30–40% of the known human transcripts that were on the array were called "present". Based on the data collected from the array, the group developed a custom-made muscle array which contains genes that are expressed in skeletal muscles [36, 37]. The Affymetrix GeneChip® HuGeneFL Array is the first version of human whole genome array and contained approximately 5000 full-length human sequences. Several different versions containing more sequences to cover the whole genome were subsequently developed, and the final/latest human version, Clariom™ D Assay, consists of more than 5,40,000 transcripts, including alternative splicing isoforms of both coding and noncoding RNAs. Microarrays for other species, including mouse, rat, nonhuman primates, insects, livestock, bird and fish, and small mammals, are also currently available for expression profiling studies. In addition to the arrays for mRNA and long noncoding RNA, GeneChip™ miRNA Arrays are available for studying small noncoding RNA, including miRNA, snoRNA, and scaRNA.

5.2.1.2 Illumina

Illumina Whole-Genome Gene Expression BeadChip consists of oligonucleotides immobilized to beads which are held in microwells on the array [38]. Up to 30 beads are available for each probe to improve data quality and reproducibility. The beads are randomly distributed across the array, one bead per well, and a 29-mer address sequence present on each bead is used for mapping the location of the beads on the array. In addition to the unique bead design, the BeadChip microarray is deployed on multi-sample array formats. Four to 24 uniform pits can be on each array, and multiple samples can be loaded to the Illumina Expression BeadChip arrays, which increases throughput and reduced sample-to-sample variations. The length of the probe is 50 bases which is synthesized in solution and then cross-linked to the beads. Labeled cRNA segments are hybridized to the probes on the BeadChip. After hybridization, washing, and staining, the image data are acquired by a scanner. The HumanHT-12 Expression BeadChip simultaneously profiles more than 47,000 transcripts representing 28,688 well-annotated genes. The transcription level is calculated using average of the signals from all the beads. In addition, some genes can be detected by more than one probe. For formalin-fixed paraffin-embedded (FFPE) samples, the Illumina developed the DASL Assay for handling degraded RNA, including muscle samples [39, 40].

5.2.1.3 Agilent Technologies

Agilent SurePrint G3 Gene Expression Microarrays are made with probes of 60 bases long [41]. The probes are synthesized onto the glass slides directly by printing A, T, C, G using an inkjet-like printer at hundreds of thousands of spots on the slide. After each nucleotide is added, a chemical de-blocking step is used to allow the next nucleotide in the chain to be added. The probes grow to the full length at the end. Agilent SurePrint used to be a two-color system Cy3 and Cy5). Now a one-color only (Cy3) system is available for the users. The advantage of this approach is that it makes updating of stock microarrays possible/easier as new gene information becomes available. In addition, custom-designed arrays targeting transcripts of specific interest can be created using this platform. The probes used here are relatively longer compared to the probes on the Affymetrix arrays but shorter than the traditional cDNA array. The length of the oligonucleotides is at a balance point for better sensitivity and specificity. Currently, human, mouse, and rat microarrays that cover coding and noncoding transcripts from the NCBI Reference Sequence (RefSeq) database are available. The coding and noncoding transcripts from RefSeq database are curated and nonredundant sequences.

5.2.2 Sequencing-Based Platforms

Next-generation sequencing (NGS)-based RNA sequencing (short for RNA-seq) is a highly sensitive and accurate method for gene expression profiling analysis that provides insight to previously undetectable changes in gene expression, as well as enabling the characterization of multiple forms of noncoding RNA. With RNA-seq, researchers can detect the various structures of the transcriptome, such as transcript isoforms, gene fusions, single nucleotide variants, and other features, without the limitation of prior knowledge. RNA-seq (1) provides sensitive, accurate measurement of gene expression at the transcript level; (2) generates both qualitative and quantitative data; (3) detects and sequences small RNAs and multiple forms of noncoding RNA, such as small interfering RNA (siRNA), microRNA (miRNA), small nucleolar (snoRNA), and transfer RNA (tRNA); (4) identifies alternatively spliced isoforms, splice sites, and allele-specific expression in a single experiment; (5) provides data sets that are not biased or restrained by existing knowledge; (6) obtains allele-specific information in the data; and (7) scales for large studies and high sample numbers. As researchers seek to understand how the transcriptome shapes biology, RNA-seq is becoming one of the most significant and powerful tools in modern science. In addition, these sequencing-based methods are more cost-effective in comparison to microarrays and real-time RT-PCR. A comparison of the major platforms is in Table 5.2 at the end of this section.

5.2.2.1 Illumina

The Illumina platforms are the most commonly used sequencers for next-generation sequencing. The technology has been used extensively for diagnosis of muscle diseases [42, 43]. It has also been used for studying muscle transcriptome to answer various biological questions, such as annotations of muscle transcripts, muscle disease mechanisms, and basic muscle biology [44–46]. Sample preparation for mRNA-seq using the Illumina platform involves isolation of RNA and chemical fragmentation of RNA, followed by reverse transcription to generate double-stranded cDNA fragments that are then sequenced. Figure 5.2 illustrates a simplified workflow of the process. Sequencing adaptors which contain barcodes are ligated to the cDNA fragments before size selection by gel electrophoresis. The fragments at desired size are then excised for sequencing. For example, fragments ranging from 250 to 400 bases are collected for regular RNA-seq. To sequence shorter fragments, e.g., to target shorter miRNA, fragments of 150 bases or less are collected. A relatively low number of cycles (9–12 cycles) of PCR amplification are performed to increase the template. Alternatively, PCR-free kits are now available to reduce biases and contamination that may be introduced during PCR amplification.

DNA libraries are hybridized to the primer lawn on the flow cell by an automated Cluster Station (cBot). Single-stranded cDNA fragments are washed across the flow cell and bind to primers on the surface of the flow cell. DNA that doesn't attach is

Table 5.2 Sequencers from Illumina, Thermo Fisher Scientific, PacBio, and Oxford Nanopore

Platform	Maximum read length (bp)	Maximum throughput per run (Gb)	Single read accuracy (%)	Strength	Weakness
Illumina					
MiSeq	2 × 300	15	99.90	Longer read length, high accuracy, lower cost	GC bias, lower output
NextSeq	2 × 150	120	99.90	High accuracy, lower cost, high throughput	GC bias
HiSeq	2 × 150	1500	99.90	High throughput, high accuracy	GC bias, short reads, high initial investment
NovaSeq	2 × 150	6000	99.90	High throughput, high accuracy	GC bias, high initial investment
ThermoFisher Scientific					
ION Torrent S5	400	25	99	Short run time, longer read length	
ION PGM	400	2	99	Short run time	Homopolymer errors
ION Proton	200	15	99	Short run time	Homopolymer errors
PacBio					
RS II	60,000	160	90	No amplification bias, long read length	Higher error rate
Oxford Nanopore					
MinION	100,000	10	90–99	No amplification bias, long read length, portable, direct detection of base modification	Higher error rate

washed away. The DNA attached to the flow cell is then replicated to form small clusters of DNA with the same sequence. This design (a cluster of the same molecule instead of only one molecule) allows florescent signals emitted during the sequencing process strong enough to be detected by a camera. During the sequencing process, the primers are first added. The DNA polymerase then adds the first fluorescently labeled terminator bases (A, C, G, and T) to the new DNA strand. Laser lights are used to activate the fluorescent label on the nucleotide base. This fluorescence is detected by a camera and recorded on a computer. Each of the terminator bases (A, C, G, and T) give off a different color. The fluorescently labeled terminator group is then removed from the first base, and the next fluorescently labeled terminator base can be added, and the process continues until the fragments are fully sequenced. Illumina sequencer sequences short reads (50 or 250 bp

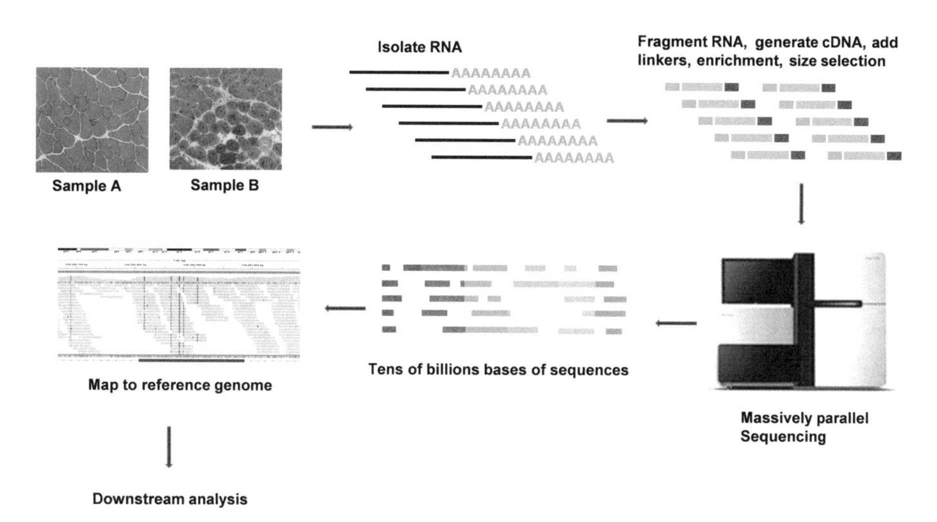

Fig. 5.2 Workflow of Illumina RNA-seq. Total RNA (or mRNA) is isolated, followed by RNA fragmentation, cDNA generation, linker ligation, fragment enrichment, and size selection to prepare library for sequencing. Next-generation sequencing is a massive parallel sequencing which generates tens of billions of bases of sequences. Downstream data analyses followed

depending on the version) and can sequence both ends of the molecules (paired ends).

5.2.2.2 Ion Torrent

The major advantage of the Ion Torrent platform is lower cost and rapid sequencing speed [47]. After reverse transcription, single-stranded DNA templates are loaded to a semiconductor chip. Unmodified A, C, G, or T dNTP are then added individually along with DNA polymerase enzyme. If an introduced dNTP is complementary to the next unpaired nucleotide of the DNA template, it will be incorporated into the complementary strand by DNA polymerase. During the incorporation process, hydrogen and pyrophosphate are released. The releasing of the hydrogen is detected by a sensor and used to determine the base at that position in the DNA template. With a portfolio of chips of varying outputs, the Ion GeneStudio™ S5 Plus and Ion PGM™ Sequencers scale to a variety of RNA-seq applications for a broad range of transcriptome sizes. For example, the Ion 550 Chip generates 100–130 million sequencing reads on the Ion GeneStudio™ S5 Plus and Prime Systems using the automated workflow of the Ion Chef System. The sequencing run time is as little as 2.5 h, with only 15 min of sample preparation time. Torrent Suite Software processes and exports the sequence reads in FASTQ or BAM formats, which can be easily imported into third-party software, such as Partek Flow software packages, for further analyses. The read length has increased from 50 bases in the original model to 400 bases in the latest model. Few examples of using this technology include studies to dissect molecular pathways involved in myogenic differentiation and de novo assembly of a muscle transcriptome [48–50].

5.2.2.3 Pacific Biosciences

PacBio uses single-molecule real-time (SMRT) sequencing technology for long-read sequencing, which allows sequencing full-length cDNA without read assembly [51]. The strengths of the system are that it is able to easily identify and quantify new transcripts and alternative splicing isoforms. The SMRT sequencing is built upon zero-mode waveguides (ZMWs) and phospholinked nucleotides. Zero-mode waveguide is an optical waveguide that guides light energy into a volume that is small in all dimensions compared to the wavelength of the light. Tens of thousands of tiny wells with ZMWs are in the SMRT cell, in which one cDNA template molecule is immobilized. Phospholinked nucleotides labeled with four different fluorophores allow observation of the addition of the nucleotides as the DNA polymerase producing the complementary strand. PacBio sequencer has advantages of long read lengths, simultaneous epigenetic characterization, and single-molecule resolution. The disadvantage is that the error rate is higher compared to the traditional short-read sequencing. Currently, muscle transcriptomic data generated using this technology were frequently used to improve annotation of an incomplete genome [52, 53].

5.2.2.4 Oxford Nanopore

Oxford Nanopore sequencing can directly sequence single molecule of DNA or RNA without the need for PCR amplification or chemical labeling of the sample. The flow cell contains more than a thousand nanopores which are nanoscale holes on a membrane. In the device, ionic currents pass through the nanoscale holes, and changes in current that occur as biological molecules pass through the nanopore are recorded by a sensor [54, 55]. Computational algorithms are used to analyze the data to reduce which of the four nucleotides passed through the pore. The approach can be used to sequence DNA, RNA, and protein. When the RNA transcripts are sequenced, the RNA samples can be reverse-transcribed into cDNA for sequencing or directly sequenced. When the RNA is sequenced directly using the technology, modifications of the bases (e.g., inosine, N^6-methyladenosine, and N^5-methylcytosine) can be detected based on the subtle differences in the current changes [56]. However, the call heavily relies and depends on the algorithms used for data analyses. While quickly improving, this platform still has the higher error rates in comparison to other major platforms. One advantage of the platform is the transportability of the smaller device, such as MinION. The pocket-size device has been brought into space and used in Antarctica as well as in rural areas for experiments and fieldwork [57–61]. This provides new opportunities for clinical diagnostic and research use. Nanopore sequencing has the potential to offer relatively low-cost sequencing, high mobility for testing, and rapid processing of samples with the ability to display results in real time. The device is the only sequencer to date that is able to sequence full-length RNAs.

5.2.3 *Reverse Transcriptase Polymerase Chain Reaction (RT-PCR) Assay*

Microarrays and next-generation sequencing are high-throughput approaches, which are great screening tools for identifying transcriptional changes in samples. The RT-PCR assay is the most commonly used method to validate the changes identified. Real-time RT-PCR allows accurate quantification because the amount of PCR amplicons in a sample is measured real time after each PCR cycle, which allows proper selection of data points for RNA quantification based on the rate of amplification. In the method, RNA is reverse-transcribed into cDNA, followed by detecting and real-time monitoring of the presence of PCR products using florescent dyes. The dyes can be either SYBR™ Green, which can be incorporated into the DNA amplicons directly, or a fluorescently labeled probe that can bind to the target sequence. In addition to individual assay, multiple assays can be performed to examine gene networks and pathways. One example is the TaqMan® Gene Expression Assay developed by Applied Biosystems. The assay utilizes the 5″ nuclease activity of the Taq DNA polymerase to cleave the fluorescently labeled probe. Each assay includes a single FAM™ dye-labeled TaqMan® probe with a minor groove binder (MGB) moiety and two unlabeled oligonucleotide primers. The assays were designed based on transcripts obtained from the NCBI Reference Sequence Project database (RefSeq). QuantStudio™ 12K Flex OpenArray® Plate allows users to specify the TaqMan® Gene Expression Assays to be included on the plate. Each plate generates 2600 data points. The advantage is the flexibility of the design which allows easy customization. The TaqMan® Array Human MicroRNA Card Set v3.0 is a two-card set containing a total of 384 TaqMan® MicroRNA Assays per card, which can be used to assay miRNAs that are differentially expressed. It was used to identify miRNAs that are differentially expressed between disease and healthy muscle cells in order to understand disease mechanisms [62].

The PCR-based technique is highly sensitive, and no pre-amplification is needed. In addition to being performed in an arrayed format for large-scale analysis, real-time RT-PCR is the gold standard technique for validating differential expressed genes identified by other high-throughput methods. Either absolute or relative quantification can be performed. Standard curve can be included to allow comparison among different plates. For validating expression changes less than twofold, digital PCR (dPCR) can be considered for their better resolution. The dPCR is a quantitative PCR method that the initial sample mix is partitioned into many individual wells prior to the PCR amplification step, resulting in either 1 or 0 targets being present in each well. Following PCR amplification, the number of positive and negative reactions is determined, and the absolute quantification of target transcript is calculated using Poisson statistics.

5.2.4 *Single-Cell RNA Sequencing*

Traditional gene expression profile analyses analyze the expression of RNAs from tissues or large populations of cells. In such mixed-cell populations, these measurements may obscure critical differences that exist between individual cells, e.g., diseased vs viable, replicating vs senescent. Single-cell RNA sequencing (scRNA-seq) allows expression profiling in individual cells. This can reveal the existence of rare cell types within a cell population that have not previously been known. In addition, scRNA-seq can be used to examine expression variations among the same type of cells. Briefly a scRNA-seq experiment involves several steps: (1) single-cell capture; (2) single-cell lysis; (3) reverse transcription of the RNA; (4) library preparation; and (5) sequencing and data analysis (Fig. 5.3). Currently, common single-cell sequencing platforms include 10X Genomics Chromium, Drop seq, and Fluidigm C1. These platforms are primarily used for single-cell capture, processing, and library preparation. A few potential applications of scRNA-seq include

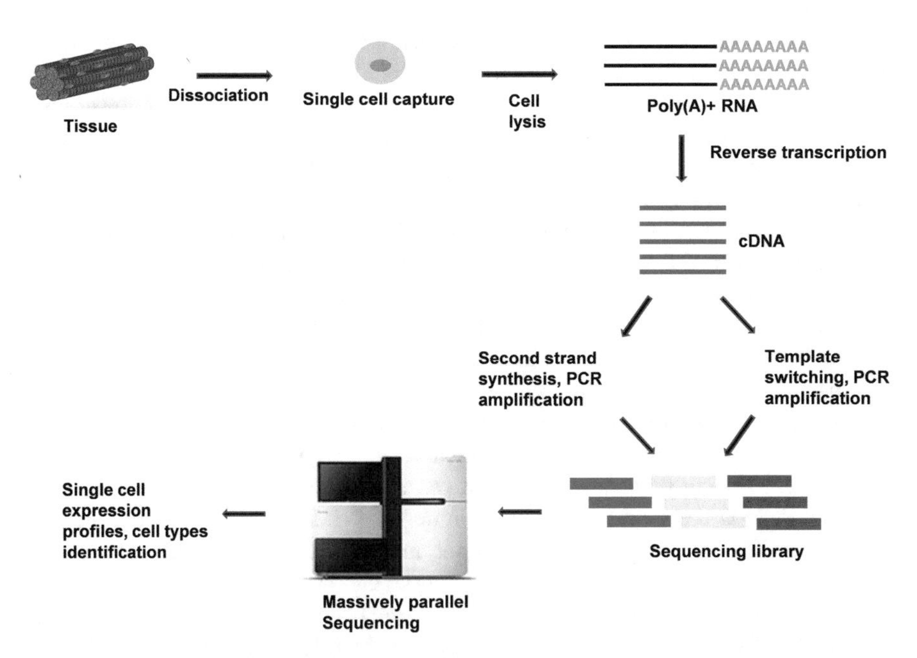

Fig. 5.3 Workflow of single-cell sequencing. Single-cell RNA sequencing starts with single-cell dissociation from the tissue sample, followed by single-cell capture. The captured cells then undergo cell lysis, reverse transcription, and fragment enrichment in preparation for sequencing and downstream data analysis

characterization of cell types, elucidation of gene regulatory networks, drug resistance clone identification, noninvasive biopsy diagnosis, stem cell lineage regulatory networks, assessing tumor heterogeneity, and CRISPR screening.

Several studies utilizing the scRNA-seq technology to study skeletal muscles suggested that muscle stem cells are a heterogeneous cell population with substantial biochemical and functional diversity [63–65]. scRNA-seq was also used to provide critical insights to disease mechanisms of the muscle disorder, such as facioscapulohumeral muscular dystrophy (FSHD). The majority of the FSHD (FSHD1) cases are caused by the contraction of a macrosatellite array called D4Z4 at the chromosome 4q35. This mutation in combination with a permissive genomic feature in the region allows the aberrant expression of double homeobox protein 4 (DUX4) protein. The aberrant expression of the DUX4 leads to downstream molecular changes that cause the disease [15, 16, 66, 67]. It has been reported that the DUX4 is not expressed in all cells and is present in only approximately 1 in 1000 proliferating myoblasts in culture [68]. Additional studies suggested that the expression of DUX4 is stochastic due to the epigenomic changes at the D4Z4 region [69, 70]. The scRNA-seq studies allowed the investigators to determine expression patterns in individual cells and conclude that a DUX4 expression induced a series of downstream expression changes [71].

5.3 Data Analyses

5.3.1 Microarrays

There are many bioinformatics tools/algorithms available to analyze microarray data. Some are command line based (mostly in R) and some are user interface (UI) based. *GeneSpring* from Agilent and Transcriptome Analysis Console (*TAC*) from Affymetrix are UI based, while *limma* and *affy* are R command line based [72, 73]. The basic analysis of microarray data can be broadly divided into steps. The first step is data extraction, preprocessing, and normalization and the second is identifying differentially expressed genes between samples, across different conditions. The *affy* package from the Bioconductor suite of tools in R extracts intensity data from .CEL files. Similarly, *GeneSpring* from Agilent Technologies helps in intensity extraction from raw microarray data sets from Agilent. The *beadArray* package from the R/Bioconductor suite of tools can be used to extract intensity values from raw images from beadArray experiments [74].

For any experiment related to the evaluation of expression differences between two groups, it is important to carry out the same experiment multiple times. This increases the sample size to perform statistical significance tests for differential expression, as well as reducing the bias in the experiment. This is typically done using one of the following methods depending on the purpose of the repeated samples. The biological replicates are samples extracted from multiple biological entities, under the same biological condition. For example, to study the effect of a drug on a muscle disease, 3 affected mice are given a drug (experimental condition) and 3 mice are not administered the drug (control). RNA is extracted from these 6 mice individually, to yield 3 biological replicates for the experiment and the control conditions. The purpose of biological replicates is to draw conclusions on the larger population of samples/controls. The technical replicates are samples extracted multiple times from the same biological entity. In this case, one mouse affected is given the drug (experiment) and the other mouse receives no drug (control). RNA is extracted once from each of the mice and RNA is separated into 3 portions each, to run 3 controls and 3 experimental samples in the microarray experiment. The purpose of technical replicates is to measure how much variation in the quantification can be expected due to technical conditions. Note technical replicates are not, and should not be, considered biological replicates; they are not used to make conclusions about the population of the mice.

In any experiment, not all replicates can be run at the same time on the same machine under the exact same conditions. Replicates may have to be run on consecutive days or on two different instruments. These conditions introduce systemic variations which, in turn, can cause variation in the expression of genes that should be identical. To quantify and account for these run-to-run variations, normalization methods are applied. There are two types of normalization, between and within arrays. Between array methods normalize the expression of genes across the multiple arrays (i.e., replicates), while within array methods normalize expression across genes within an array (within the same replicate). One of the most commonly used methods of between array normalization is quantile normalization [75]. The *RMA* function in the *affy* package, part of the Bioconductor suite of tools and GeneSpring from Agilent technology, not only performs quantile normalization but also performs background correction, probe-level intensity calculation, and probe set summarization. For within array normalization, Loess normalization is typically used and implemented by the *limma* package, part of the Bioconductor suite of tools. Other platforms like nimblegen and illumine have the packages *oligo* [76] and *beadArray*, respectively, which extract intensity and normalize data using the RMA function.

To evaluate possible changes in expression pattern across replicates, it is important to use visualization in combination with statistical methods. A useful visualization method is the *heatmap.2* function in the *gplots* R package [77], which combines heatmap functionality with hierarchical clustering to visualize the expression patterns across replicates. Ideally, a normalized data set should show a similar expression pattern for a given gene across multiple replicates. If a significant difference in expression patterns is observed, one needs to first verify that the correct file was used for a given replicate, an easy mistake to make given the similarity in file names. Next, different normalization methods should be used to verify that they show the same general pattern. Principal component analysis (PCA) [77] and box plots are used by *Genespring* to visualize the expression patterns across samples. This normalization allows one to be confident that the quantification of gene expression from different samples can be directly compared when assessing differences in gene expression affected by an introduced factor (i.e., treatment, disease, etc.).

When comparing two or more groups, one can identify genes that are differentially expressed using statistical analyses. Difference in expression is expressed as fold change (experimental gene expression/control gene expression) or logarithmic fold change [$\log_2$ (experimental gene expression/control gene expression)]. Statistical significance is measured using hypothesis testing based on the design of the experiment. See the chapter of bioinformatics and statistics for omics data for details on the appropriate statistical method to use. Often, for designs where there is a single experimental condition and a single control condition, Student's t-test is used [78], whereas ANOVA is commonly used for complex designs where there are multiple experimental and control conditions [79]. Be advised however that both Student's t-test and ANOVA have assumptions that the data must meet for the tests to be valid. Multiple testing correction methods such as the Benjamini-Hochberg [80] or Bonferroni [81] are used to limit false positives. Base R packages are available to perform t-tests (*t.test()*), ANOVAs (*aov()*), and multiple testing p-value correction (*p.adjust*). GeneSpring uses ANOVA in their differential expression analysis. Another program, limma, is a Bioconductor package which is based on linear regression models and is one of the most popular differential expression analysis tools. Before beginning the analysis of differential expression, a clear understanding of the appropriate statistical method to use is essential, as using the incorrect method can lead to invalid conclusions. Once genes are found to be differentially expressed between conditions, they can be further analyzed by various means including clustering analyses and functional annotation and gene ontology methods. In addition, Ingenuity Pathway Analyses, a web-based commercial tool, can be used to identify networks, pathways, and regulatory relationships among the identified transcripts.

5.3.2 RNA-Sequencing (RNA-Seq)

High-throughput sequencing techniques produce either short (50–75 bp) reads like those generally produced by Illumina sequencers or longer reads produced by newer methods such as PacBio (30–50kbp) [82] and Oxford Nanopore (can be longer than 100K) [55, 83] systems. Read length is an important parameter, as longer reads are better able to estimate the number of counts of larger genes. Another important parameter is the coverage of reads across a gene. To get a greater coverage, the reads are sequenced from both ends (also known as paired end reads) when the Illumina platform is used. Although the read lengths are different, major workflow is similar and described below.

The major workflow, from raw sequences obtained from the sequencers (here we specifically discuss short-read sequencers) to a list of differentially expressed genes, can be divided into four major steps (workflow is depicted visually in Fig. 5.4):

Step 1 is preprocessing of raw sequences. This step uses the same tools as used in Quality Control in Genomics analysis section.

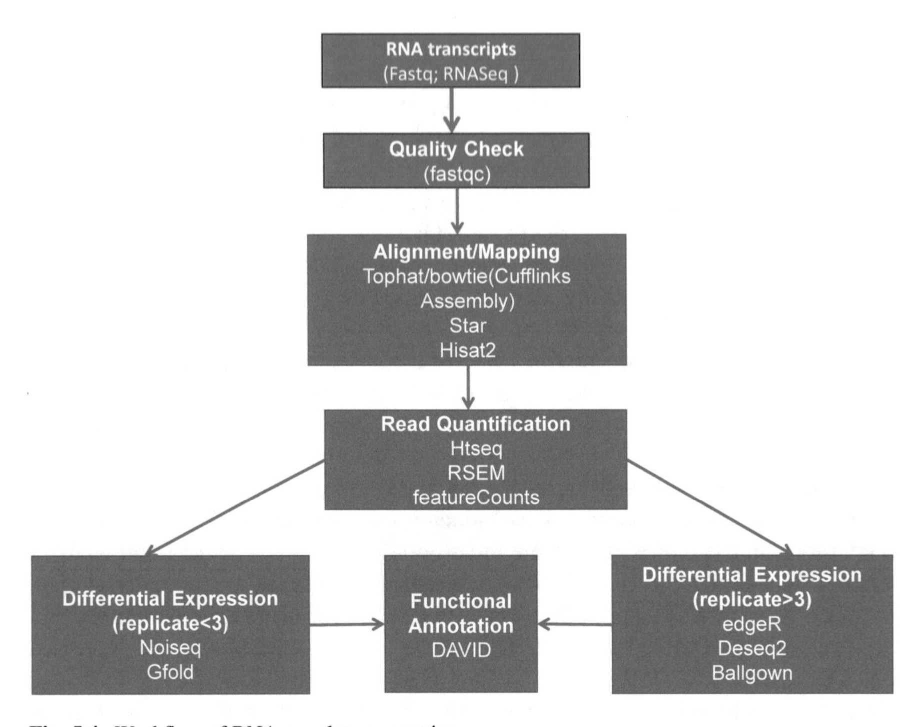

Fig. 5.4 Workflow of RNA-seq data processing

Step 2 is aligning the sequence. Multiple tools/algorithms are available for the alignment/mapping of the transcriptome sequence to the reference genome. The important aspects in choosing an alignment algorithm/software are the accuracy of alignment to the reference genome, computational memory used, and the time taken for the alignment. In addition, for transcriptome alignments, it is important that the aligners are splice junction aware. This is because the RNA-seq is performed on mature messenger RNAs (mRNAs), which are devoid of introns, but the reference genome contains the intron sequences. If the aligner is not able to accurately identify the region where splicing or removal of introns (splicing junction) happens, then it would treat it as a deletion, and not estimate the transcript count accurately. *Tophat2* [84], part of the Tuxedo [85] suite of tools, is a read aligner that is splice junction aware. The Tuxedo suite of tools takes a raw FASTQ file as an input, aligns it to the reference genome (*Tophat2*), and assembles it to accurately calculate expression. *Cufflinks* [86] calculates the differential expression between samples (*cuffdiff*) and finally visualizes expression plots (*Cummerbund*). *Tophat2* uses *Bowtie2* [87] (a fast aligner that uses Burrows-Wheeler transform (BWT) [88] for compression and storage of the reference genome) and FM index [89] (a compressed indexing method for Burrows-Wheeler transformation), so that it can be accessed rapidly. The regions that do not align to the reference genome using bowtie are divided into smaller segments using *Tophat2*. *Tophat2* determines splice junction, when it observes read segments aligning to the reference genome with a gap of 100–1000 bases between them.

STAR (spliced transcripts alignment to a reference) [90] is a fast aligner which uses the concept of Maximal Mappable Prefix (MMP), on uncompressed suffix arrays. Suffix arrays are a type of data structure, which enables faster searching of sequences for alignment, by breaking the larger sequence into smaller subsections (suffix), sorting and storing the address in the form of arrays. The MMP algorithm searches these arrays to find the longest subsection of sequence reads that map to the reference genome sequence. After the alignment, the software performs clustering and stitching of all reads aligned to the genome to form a complete realigned sequence. The disadvantage of this algorithm is that it is memory intensive as the MMP search algorithm is performed on an uncompressed suffix array instead of a compressed one. *HiSat* [91], a newer alignment algorithm, is part of the new Tuxedo 2 [92] pipeline [*HiSat* (alignment), *StringTie* (assembly) [93], and *Ballgown* (differential expression calculation) [94]]. Like Tophat2, *HiSat* also uses bowtie for alignment utilizing the FM index functionality but also implements two different FM indexes that increase the accuracy of alignment. Of the FM indices, one is a global FM index and comprises the whole reference genome. The other index contains several small FM indices, each made up of a 64,00-bp subsection, which together covers the whole genomic region. By using memory optimization algorithms, it can reduce memory usage and can align a whole human genome using only 4GB of memory.

Step 3 is assembly and quantification. The output from the alignment processes produces aligned reads in the Sequence Alignment Map (SAM) format [95]. The SAM file is text-based and includes mandatory fields of chromosome number, location on the chromosome, and quality annotated, i.e., no genes are associated to the reads. The assembly and quantification tools are responsible for annotating the reads and estimating read counts associated with each transcript in the genome. SAM files are converted into BAM files (binary SAM files), sorted, indexed, and then given as an input to the assembler or read count estimation tools. Quality check of the BAM files can be done using *RSeQC* [96]. Preprocessing is performed by *samtools* software. There are multiple processes that perform assembly and quantification. The choice of read count quantification tool depends on whether one wants to have normalized read counts or raw read counts that can be normalized using custom scripts.

Normalized read counts can be divided into two basic types, reads per kilo million bases (RPKM) and transcripts per kilobase millions (TPM). RPKM is a normalized read count for single-read RNA sequences [97]. It is measured as the ratio of the number of reads depicting a region and the product of total reads divided by 1 million and region length divided by 1000. Fragments per kilo million (FPKM) bases is the paired end interpretation of RPKM. One of the tools that produce FPKM values is *Cufflinks*. *Cufflinks*, part of the Tuxedo pipeline, assembles the reads by first creating an overlap graph which represents all the reads that map to a region. The algorithm then traverses through the graph to assemble the isoforms, by identifying the minimum number of transcripts that can signify a particular intron junction. The assembled transcript fragments (transfrags) are counted using a statistical model for RNA-seq experiments [86]. The output from this tool is a Gene transfer format (GTF) file, which, along with information like chromosome name and location of the gene, also contains the gene abundance or read counts in FPKM. TPM is another normalization method [98] that is measured as the ratio of reads per kilobase (RPK) per million scaling factors. The RPK value is the read counts of each gene divided by the length of each gene in kilobases. The per million scaling factor is the sum of all RPK in a sample divided by 10,00,000. As the denominator remains constant, TPM for a given gene remains constant across replicate conditions if the same number of reads is aligned to the reference genome. This is not the case for RPKM (or FPKM) as its denominator is the length of the genes, which is a variable factor.

RSEM tool is responsible for the calculation of TPM for reads [99]. In the workflow, *rsem-prepare-reference* function takes in whole genome reference sequence in fasta format and GTF files and converts it to a transcript reference sequence. RSEM utilizes the transcript reference to get a clear count of isoforms. Next, *rsem-calculate-expression* calculates the TPM read counts. The input to this tool can be either FASTQ files or aligned BAM files. If fastq is provided as an input, the function first aligns the reads to a reference genome (default aligner is bowtie, STAR is the other aligner available in RSEM). Then using the expectation-maximization algorithm [100], maximum likelihood abundance (expected counts) is calculated and converted into TPM values. In case of aligned BAM files, special caution must be taken, if they are produced by aligners other than BWA and STAR. Because RSEM uses an enhanced read generation model for estimating read abundance compared to other aligners, aligner parameters must be changed to report all aligned reads. Outputs from this tool are 2 text files reporting gene TPM and isoform TPM values, respectively.

There are few other tools that do not perform any normalization in the read estimation step. These provide users with raw read counts which can then be normalized either by using available R/Python scripts or by functions available in downstream processes. One of the tools, Htseq a python package [101], not only quantifies read counts but can also be used as a parser for different genomic data files like FASTQ, SAM/BAM, a quality assessment tool (*htseq-qa*) for aligned reads, and GenomicArray class that stores the information from genomic data. The *htseq-count* function counts the reads based on the idea that the reads should cover an exon either completely or partially. The read counts are raw read counts, which can be normalized into FPKM, by custom scripts in python or R. Other useful tools are *featureCounts* [102] and summarizeOverlaps of the *Genomic Alignment* package in R [103].

Step 5 is differential expression calculation to identify genes or noncoding RNAs that are statistically significantly different between two or more conditions. In these steps raw or normalized reads are taken as an input, and using either parametric or nonparametric methods, differentially expressed genes are evaluated. The reads are given as an input directly as text files (in the case of Cuffdiff) or as R data objects extracted by tools like *tximport* [104], for downstream processing with R Bioconductor packages. Sample size is an important criterion to identify truly statistically significant differentially expressed genes. For RNA-seq tools, the efficiency of detection increases with sample size, and most of the statistical tools work reasonably well with a sample size of 3–5 per condition. Many of these tools can identify statistically significant targets with no replicates; however, extreme caution should be taken with interpretation. Please see the chapter of bioinformatics and statistics for omics data to fully understand the appropriate sample size and statistical methods to use. In general, differential expression tools can be divided into two types.

The first one is parametric analysis. The read counts generated from the quantification tools are either raw read counts or normalized read counts (FPKM/TPM). Most of the downstream differential expression tools have methods of normalization built into the packages. *EdgeR* [105], a R Bioconductor package, takes as input the read counts and stores them in a list-based data object called DGEList. This contains the read count matrix and the sample information, including the library information, in the form of a R type data frame and an optional data frame containing annotated gene information. *EdgeR* has methods to filter data based on count per million (CPM; which is a normalization method similar to TPM) read counts, to normalize data (*calcNormFactors*), and to estimate dispersion in read counts across samples using a quantile-adjusted conditional maximum likelihood negative binomial model (*estimateDisp*). For the differential expression calculation, a design matrix containing the condition and sample information (i.e., which samples are controls and which are experimental) must be provided as an input. Differential expression is calculated using an exact negative binomial test (*exactTest*). Differentially expressed genes are listed with parameters showing statistical significance (p-value and adjusted p-value) and fold change by the topTags function. *Deseq2* is another Bioconductor package that uses similar statistics as *EdgeR* [106]. It takes as input raw, non-normalized read counts and stores them as a DESeqDataSet, similar to DGEList for EdgeR. Instead of three different functions to calculate differential expression, Deseq2 has the function *DESeq* which contains the functions for normalization, dispersion estimation, and negative binomial generalized linear model (GLM) [107] fitting followed by a Wald test [108] to test for differential expression. This tool provides an output similar to *EdgeR*. *Cuffdiff*, part of the Tuxedo suite of tools, uses a merged GTF file produced by *Cufflink* along with aligned SAM files from the conditions (output from *Tophat*) as the input for finding differential expression between conditions. The GTF files for each of the samples produced by *Cufflinks* are merged using the *Cuffmerge* function. The *Cuffdiff* function works twofold: First, it quantifies the reads to FPKM values. Second, it performs a statistical significance test based on negative binomial distribution as the previous two functions. The output is quite similar to the previous two methods.

The second is nonparametric analysis. Although parametric methods are quite powerful, it is completely dependent on the assumption that the distribution of the expression values fit the distribution of the statistical test being used. Methods like *GFold* [109] and *Noiseq* [110] use nonparametric methods of differential expression calculation. *GFold* uses a posterior distribution of logarithmic value of the raw fold changes and then ranks them based on their values. Those with higher ranks are upregulated, whereas low-ranked ones are downregulated. GFold can be used for both single replicates and multiple replicates. For *Noiseq*, several distributions are created; a noise distribution is for exhibiting change in counts, a contrasting fold change distribution (M), and an absolute expression difference (D), for each gene in a given condition. These are then used to compare whether M and D values between conditions fall within the noise or are actual differences. NoiSeq has 2 functions, *NOISeq-real* for replicates and *NOISeq-sim* for non-replicated samples. The output from these steps is a list of genes exhibiting statistically significantly different expressions between conditions. This list is used in further downstream analysis as described in the chapter of bioinformatics and statistics for omics data.

5.3.3 Single-Cell Sequencing

Different cells in an organism exhibit different phenotypic functionalities. These functionalities are governed by their genetic architecture; thus, it is important to understand the expression of genes in individual cells. The single-cell sequencing technique uses NGS methods on RNA isolated from single cells [111]. A review of library preparation and sequencing technique for single cell can be found in [112]. The bioinformatics pipeline is similar to RNA-seq, wherein quality check is done by FastQC, alignment performed by *STAR*, followed by read count estimation by *HTSEq*, and differential expression analysis performed by *Deseq2*. Custom-made codes to differentiate cells based on barcodes are present in pipelines developed by companies that develop the libraries such as 10Xgenomics. Identification of different cell types is done using PCA or t-distributed stochastic neighbor embedding (t-SNE) [113]. These assist in differentiation and visualization of gene clusters based on expression similarities. For example, genes having similar differential expression patterns among conditions would be clustered together and by identifying markers we can identify known cell types or unknown cell types. T-SNE is implemented in *Seurat* [114].

5.4 Platform Selection and Limitations

Gene expression profiles of different tissues, cells, conditions, disease states, or even single cells are now routinely generated. This explosion in gene expression profiling has been deeply affected by the rapid development of new technologies with an improved sensitivity and cost effect. Which platform to be used for a study depends on the goal of the study. The advantages of next-generation sequencing are that the method provides direct counts of molecules; it can be used to study all types of RNA transcripts; it provides better resolution for transcripts that share similar sequences; alternative splice isoforms can be directly detected; and it can be used to examine allele-specific expression. The disadvantages of the NGS approach are that it may not be able to detect low-abundance transcripts; data analyses and statistical analyses are more challenging, and the final results heavily rely on how the raw data were processed; computational demands are high; data storage and sharing are challenging; and there are concerns of privacy when sequence information is examined. In addition to the above, the cost and availability of the platform and bioinformatics expertise are something to be considered.

To select a proper platform and plan a profiling study, which RNA population will be studied is an important factor. NGS will be the choice if whole transcriptome, including those that are not on the microarrays, is studied. If one is only interested in a specific group of RNA transcripts or few specific pathways, it may be more cost-effective to use arrays and PCR-based assays. In addition to commercially available stock arrays and PCR panels, researchers can custom-design arrays and PCR-based assays to include specific transcripts of interests. The transcripts are well annotated, and data analyses process is straightforward; therefore, the turnaround time is shorter. Another factor to be considered is that NGS is limited by its total read counts for each run. For example, the highly abundant rRNAs need to be removed from the samples to increase the reads of the rest of the transcripts. The transcripts that are low in abundance are often missed by direct sequencing. These include low-abundant mRNA transcripts, alternatively spliced variants, and lncRNA transcripts. In general, 20–50 million reads are sufficient for detecting ~20,000 transcripts. 100 million reads will increase coverage and allow proper quantification and identify differentially expressed transcripts. Three hundred million reads give enough depth for studying alternative splicing and higher number of reads may be needed for studying lncRNA. Since commercial microarrays that cover the whole transcriptome, noncoding RNA, and splice variants are available, one can decide which platform to use based on the experimental aims, turnaround time, and costs. While the microarrays are generally good at detecting low-abundant transcripts, depending on the dynamic range of different microarray platforms, sensitivities to changes of highly abundant transcripts may reduce due to saturation of the intensity signals at the high end.

One of the important considerations when conducting the profiling studies is normalization of the data. The method used for normalization and for hybridization-based assays is usually less an issue when a large number of genes are examined at the same time. For example, Affymetrix normalizes the intensity of individual data point to the average intensity of the whole array. This strategy is based on the assumption that most of the genes in a sample do not change significantly. Any conditions that change the expression of a large number of genes toward the same direction will not be suitable to use this normalization method. Instead, one can identify genes that express at consistent level in all samples to be normalized to. The same strategies can be used when the platform does not examine a large number of genes, such as cDNA arrays or PCR arrays. For muscle research, commonly used internal controls such as glyceraldehyde-3-phosphate dehydrogenase (GAPDH), β-actin, and 18S rRNA may have inconsistent expression levels in samples that are severely affected by a condition or disease. An important question is whether the condition or disease you are studying will cause genome-wide transcriptional changes toward one direction or changes of the gene level of the internal controls.

Increased heterogeneity of pathological specimen needs to be kept in mind when interpreting transcriptomic data. In addition to degeneration/regeneration, a muscle sample may contain different degrees of pathological changes such as inflammation, fibrosis, and fatty tissue replacement. One can imagine that severe fat infiltration or inflammation can completely change the profiles. When the pathological changes are prominent, we are comparing different tissues in such situations. This will cause a challenge when analyzing data as well as interpreting results. To answer the questions regarding the origin of the changes, additional studies are needed, such as immunostaining or immunohistochemistry to visualize and locate the protein products of the genes that were shown to change in the profiling data [6, 7, 18]. Now one can also use single-cell sequencing to examine gene profiles in each cell individually and determine which cells contribute to the changes [115]. However, it is still a good idea to validate the protein changes in the tissue directly.

While transcriptomic approaches can potentially generate large amount of data, usually only a limited number of genes and pathway are reported in a publication. This can be due to (1) insufficient statistical power due to small sample sizes; (2) lack of bioinformatics expertise to fully analyze the data; and (3) the researcher selecting and focusing on only part of the genes and pathways to follow up. Placing expression profiling results in a publicly accessible microarray database makes it possible for other researchers to access the data and have new discoveries beyond the scope of published results [116].

5.5 Vital Future Directions

Technologies for studying transcriptome have been evolving quickly. New methods and approaches to address common issues and concerns associated with the current approach are becoming available. Companies have been improving reagents for better sample preparation and producing new instruments for higher throughput and lower cost per base. In addition, new utilization and approaches are developed by both the companies and users to answer specific questions. Here we discuss few examples. One of the questions is issues associated with tissue heterogeneity in muscle samples. scRNA-seq allows single-cell resolution of transcriptomes; however, the information on the location of the cells and the relationship among the cells are lost during the sample processing. To allow researchers to learn not only what is in a cell but how the cells interact with other cells provides an invaluable insight into understanding muscle biology and disease mechanisms. Recently 10X Genomics acquired Spatial Transcriptomics which provides such technology. The technology for the spatial gene expression profiling was originally developed at Science for Life Laboratory in Stockholm, Sweden, which allows RNA sequencing to be performed from tissue sections [117]. The process includes first attaching a frozen-sectioned tissue section to a specialized chip. The tissue section is imaged before the RNA transcripts in the tissue section are reverse-transcribed. The image with coordinate information will later be used to be matched to the gene expression data generated from each specific location. The chip contains an array of probes which have poly-T tails at the end. After the tissue section is fixated and permeabilized, the RNA with poly-A sequence will be captured by the probes and reverse-transcribed. The cDNA-RNA hybrids are cleaved off the chip, followed by library preparation for RNA-seq. The RNA-seq data are integrated with the histology data for visualizing the transcriptomes of cells at different locations in the section.

One challenge in performing single-cell profiling on skeletal muscle samples is that the myofibers are multinucleated cells. It is not feasible to capture individual myofiber for profiling using current platforms. On the other hand, single-nucleus profiling will provide transcriptome information of individual nucleus in a sample. A recent study showed that comparable results were obtained by using scRNA-seq and snRNA-seq approaches although the snRNA-seq is better for sequencing nucleus-enriched lncRNAs and miRNA precursors [64].

In addition to sequences, new technologies allow researchers to study RNA base modifications [118]. More than a hundred RNA base modifications are known but the role of the modifications in mRNA is mostly unclear. The Oxford Nanopore platform allows direct sequencing of RNA molecules and identifying RNA modifications. The epitranscriptome will be one future direction that can take advantage of the new technologies.

Transcriptomic approaches allow the researcher to determine the transcription activities of all active genes. Temporal profiling approaches use a series of profiles obtained at different time points to construct temporal changes of gene activities over a period of time. Meanwhile, researches combine more than one omics approaches to understand how the other processes, such as epigenomic, posttranscriptional, and posttranslational regulations, affect the cell functions and to gain a more integrated picture. Using multiple omics approaches to obtain temporal data makes construction of comprehensive molecular pathways involved in specific conditions possible. Data from single-cell profiling and spatial gene profiling increases resolution to changes in individual cell.

References

1. Li, Z., et al. (2018). Systematic transcriptome-wide analysis of mRNA-miRNA interactions reveals the involvement of miR-142-5p and its target (FOXO3) in skeletal muscle growth in chickens. *Molecular Genetics and Genomics: MGG, 293*, 69–80. https://doi.org/10.1007/s00438-017-1364-7.
2. Li, R., et al. (2019). Characterization and expression profiles of muscle transcriptome in Schizothoracine fish, *Schizothorax prenanti*. *Gene, 685*, 156–163. https://doi.org/10.1016/j.gene.2018.10.070.
3. Cote, L. E., Simental, E., & Reddien, P. W. (2019). Muscle functions as a connective tissue and source of extracellular matrix in planarians. *Nature Communications, 10*, 1592. https://doi.org/10.1038/s41467-019-09539-6.

4. Burniston, J. G., et al. (2013). Gene expression profiling of gastrocnemius of "minimuscle" mice. *Physiological Genomics, 45*, 228–236. https://doi.org/10.1152/physiolgenomics.00149.2012. physiolgenomics.00149.2012 [pii].
5. Pietu, G., et al. (1996). Novel gene transcripts preferentially expressed in human muscles revealed by quantitative hybridization of a high density cDNA array. *Genome Research, 6*, 492–503.
6. Chen, Y. W., Hubal, M. J., Hoffman, E. P., Thompson, P. D., & Clarkson, P. M. (2003). Molecular responses of human muscle to eccentric exercise. *Journal of Applied Physiology, 95*, 2485–2494.
7. Chen, Y. W., et al. (2002). Response of rat muscle to acute resistance exercise defined by transcriptional and translational profiling. *The Journal of Physiology, 545*, 27–41.
8. Bonafiglia, J. T., Menzies, K. J., & Gurd, B. J. (2019). Gene expression variability in human skeletal muscle transcriptome responses to acute resistance exercise. *Experimental Physiology, 104*, 625–629. https://doi.org/10.1113/EP087436.
9. Turner, D. C., Seaborne, R. A., & Sharples, A. P. (2019). Comparative transcriptome and methylome analysis in human skeletal muscle anabolism, hypertrophy and epigenetic memory. *Scientific Reports, 9*, 4251. https://doi.org/10.1038/s41598-019-40787-0.
10. Dickinson, J. M., et al. (2018). Transcriptome response of human skeletal muscle to divergent exercise stimuli. *Journal of Applied Physiology (1985), 124*, 1529–1540. https://doi.org/10.1152/japplphysiol.00014.2018.
11. Mahmassani, Z. S., et al. (2019). Age-dependent skeletal muscle transcriptome response to bed rest-induced atrophy. *Journal of Applied Physiology (1985), 126*, 894–902. https://doi.org/10.1152/japplphysiol.00811.2018.
12. Vechin, F. C., et al. (2019). Low-intensity resistance training with partial blood flow restriction and high-intensity resistance training induce similar changes in skeletal muscle transcriptome in elderly humans. *Applied Physiology, Nutrition, and Metabolism = Physiologie appliquee, nutrition et metabolisme, 44*, 216–220. https://doi.org/10.1139/apnm-2018-0146.
13. Chen, Y. W., et al. (2005). Early onset of inflammation and later involvement of TGFbeta in Duchenne muscular dystrophy. *Neurology, 65*, 826–834.
14. Dadgar, S., et al. (2014). Asynchronous remodeling is a driver of failed regeneration in Duchenne muscular dystrophy. *The Journal of Cell Biology, 207*, 139–158. https://doi.org/10.1083/jcb.201402079. jcb.201402079 [pii].
15. Sharma, V., Harafuji, N., Belayew, A., & Chen, Y. W. (2013). DUX4 differentially regulates transcriptomes of human rhabdomyosarcoma and mouse C2C12 cells. *PLoS One, 8*, e64691. https://doi.org/10.1371/journal.pone.0064691. PONE-D-13-08552 [pii].
16. Dixit, M., et al. (2007). DUX4, a candidate gene of facioscapulohumeral muscular dystrophy, encodes a transcriptional activator of PITX1. *Proceedings of the National Academy of Sciences of the United States of America, 104*, 18157–18162.
17. Wang, E. T., et al. (2019). Transcriptome alterations in myotonic dystrophy skeletal muscle and heart. *Human Molecular Genetics, 28*, 1312–1321. https://doi.org/10.1093/hmg/ddy432.
18. Chen, Y. W., Zhao, P., Borup, R., & Hoffman, E. P. (2000). Expression profiling in the muscular dystrophies: Identification of novel aspects of molecular pathophysiology. *The Journal of Cell Biology, 151*, 1321–1336.
19. Zhang, N., et al. (2018). Dynamic transcriptome profile in db/db skeletal muscle reveal critical roles for long noncoding RNA regulator. *The International Journal of Biochemistry and Cell Biology, 104*, 14–24. https://doi.org/10.1016/j.biocel.2018.08.013.
20. Scott, L. J., et al. (2016). The genetic regulatory signature of type 2 diabetes in human skeletal muscle. *Nature Communications, 7*, 11764. https://doi.org/10.1038/ncomms11764.
21. Gallagher, I. J., et al. (2012). Suppression of skeletal muscle turnover in cancer cachexia: Evidence from the transcriptome in sequential human muscle biopsies. *Clinical Cancer Research: An Official Journal of the American Association for Cancer Research, 18*, 2817–2827. https://doi.org/10.1158/1078-0432.CCR-11-2133.

22. Chen, Y. W., et al. (2017). Molecular signatures of differential responses to exercise trainings during rehabilitation. *Biomedical Genetics and Genomics, 2*. https://doi.org/10.15761/BGG.1000127.
23. Boehler, J. F., et al. (2017). Effect of endurance exercise on microRNAs in myositis skeletal muscle—A randomized controlled study. *PLoS One, 12*, e0183292. https://doi.org/10.1371/journal.pone.0183292.
24. Benoit, B., et al. (2017). Fibroblast growth factor 19 regulates skeletal muscle mass and ameliorates muscle wasting in mice. *Nature Medicine, 23*, 990–996. https://doi.org/10.1038/nm.4363.
25. Wu, J., et al. (2014). Ribogenomics: The science and knowledge of RNA. *Genomics, Proteomics and Bioinformatics, 12*, 57–63. https://doi.org/10.1016/j.gpb.2014.04.002.
26. Varemo, L., et al. (2016). Proteome- and transcriptome-driven reconstruction of the human myocyte metabolic network and its use for identification of markers for diabetes. *Cell Reports, 14*, 1567. https://doi.org/10.1016/j.celrep.2016.01.054.
27. Lundberg, E., et al. (2010). Defining the transcriptome and proteome in three functionally different human cell lines. *Molecular Systems Biology, 6*, 450. https://doi.org/10.1038/msb.2010.106.
28. Vogel, C., & Marcotte, E. M. (2012). Insights into the regulation of protein abundance from proteomic and transcriptomic analyses. *Nature Reviews. Genetics, 13*, 227–232. https://doi.org/10.1038/nrg3185.
29. Nie, L., Wu, G., Brockman, F. J., & Zhang, W. (2006). Integrated analysis of transcriptomic and proteomic data of Desulfovibrio vulgaris: Zero-inflated Poisson regression models to predict abundance of undetected proteins. *Bioinformatics, 22*, 1641–1647. https://doi.org/10.1093/bioinformatics/btl134.
30. Greenbaum, D., Colangelo, C., Williams, K., & Gerstein, M. (2003). Comparing protein abundance and mRNA expression levels on a genomic scale. *Genome Biology, 4*, 117. https://doi.org/10.1186/gb-2003-4-9-117.
31. Margulies, M., et al. (2005). Genome sequencing in microfabricated high-density picolitre reactors. *Nature, 437*, 376–380. https://doi.org/10.1038/nature03959.
32. Heather, J. M., & Chain, B. (2016). The sequence of sequencers: The history of sequencing DNA. *Genomics, 107*, 1–8. https://doi.org/10.1016/j.ygeno.2015.11.003.
33. Schena, M., Shalon, D., Davis, R. W., & Brown, P. O. (1995). Quantitative monitoring of gene expression patterns with a complementary DNA microarray. *Science, 270*, 467–470.
34. DeRisi, J., et al. (1996). Use of a cDNA microarray to analyse gene expression patterns in human cancer. *Nature Genetics, 14*, 457–460. https://doi.org/10.1038/ng1296-457.
35. Dalma-Weiszhausz, D. D., Warrington, J., Tanimoto, E. Y., & Miyada, C. G. (2006). The affymetrix GeneChip platform: An overview. *Methods in Enzymology, 410*, 3–28. https://doi.org/10.1016/S0076-6879(06)10001-4.
36. Kostek, M. C., et al. (2007). Gene expression responses over 24 h to lengthening and shortening contractions in human muscle: Major changes in CSRP3, MUSTN1, SIX1, and FBXO32. *Physiological Genomics, 31*, 42–52.
37. Borup, R. H., et al. (2002). Development and production of an oligonucleotide MuscleChip: Use for validation of ambiguous ESTs. *BMC Bioinformatics, 3*, 33.
38. Fan, J. B., et al. (2006). Illumina universal bead arrays. *Methods in Enzymology, 410*, 57–73. https://doi.org/10.1016/S0076-6879(06)10003-8.
39. Kotorashvili, A., et al. (2012). Effective DNA/RNA co-extraction for analysis of microRNAs, mRNAs, and genomic DNA from formalin-fixed paraffin-embedded specimens. *PLoS One, 7*, e34683. https://doi.org/10.1371/journal.pone.0034683.
40. Kibriya, M. G., et al. (2010). Analyses and interpretation of whole-genome gene expression from formalin-fixed paraffin-embedded tissue: An illustration with breast cancer tissues. *BMC Genomics, 11*, 622. https://doi.org/10.1186/1471-2164-11-622.

41. Wolber, P. K., Collins, P. J., Lucas, A. B., De Witte, A., & Shannon, K. W. (2006). The agilent in situ-synthesized microarray platform. *Methods in Enzymology, 410*, 28–57. https://doi.org/10.1016/S0076-6879(06)10002-6.
42. Wu, L., Brady, L., Shoffner, J., & Tarnopolsky, M. A. (2018). Next-generation sequencing to diagnose muscular dystrophy, rhabdomyolysis, and HyperCKemia. *The Canadian Journal of Neurological Sciences. Le journal canadien des sciences neurologiques, 45*, 262–268. https://doi.org/10.1017/cjn.2017.286.
43. Nigro, V., & Piluso, G. (2012). Next generation sequencing (NGS) strategies for the genetic testing of myopathies. *Acta myologica: Myopathies and Cardiomyopathies: Official Journal of the Mediterranean Society of Myology, 31*, 196–200.
44. Hestand, M. S., et al. (2010). Tissue-specific transcript annotation and expression profiling with complementary next-generation sequencing technologies. *Nucleic Acids Research, 38*, e165. https://doi.org/10.1093/nar/gkq602.
45. Colangelo, V., et al. (2014). Next-generation sequencing analysis of miRNA expression in control and FSHD myogenesis. *PLoS One, 9*, e108411. https://doi.org/10.1371/journal.pone.0108411.
46. Cardoso, T. F., et al. (2017). RNA-seq based detection of differentially expressed genes in the skeletal muscle of Duroc pigs with distinct lipid profiles. *Scientific Reports, 7*, 40005. https://doi.org/10.1038/srep40005.
47. Pennisi, E. (2010). Genomics. Semiconductors inspire new sequencing technologies. *Science, 327*, 1190. https://doi.org/10.1126/science.327.5970.1190.
48. Tripathi, A. K., et al. (2014). Transcriptomic dissection of myogenic differentiation signature in caprine by RNA-Seq. *Mechanisms of Development, 132*, 79–92. https://doi.org/10.1016/j.mod.2014.01.001.
49. Parmakelis, A., Kotsakiozi, P., Kontos, C. K., Adamopoulos, P. G., & Scorilas, A. (2017). The transcriptome of a "sleeping" invader: De novo assembly and annotation of the transcriptome of aestivating Cornu aspersum. *BMC Genomics, 18*, 491. https://doi.org/10.1186/s12864-017-3885-1.
50. Possidonio, A. C., et al. (2014). Cholesterol depletion induces transcriptional changes during skeletal muscle differentiation. *BMC Genomics, 15*, 544. https://doi.org/10.1186/1471-2164-15-544.
51. Cartolano, M., Huettel, B., Hartwig, B., Reinhardt, R., & Schneeberger, K. (2016). cDNA Library Enrichment of Full Length Transcripts for SMRT Long Read Sequencing. *PLoS One, 11*, e0157779. https://doi.org/10.1371/journal.pone.0157779.
52. Chen, S. Y., Deng, F., Jia, X., Li, C., & Lai, S. J. (2017). A transcriptome atlas of rabbit revealed by PacBio single-molecule long-read sequencing. *Scientific Reports, 7*, 7648. https://doi.org/10.1038/s41598-017-08138-z.
53. Masonbrink, R. E., et al. (2019). An annotated genome for *Haliotis rufescens* (Red Abalone) and resequenced green, pink, pinto, black, and white abalone species. *Genome Biology and Evolution, 11*, 431–438. https://doi.org/10.1093/gbe/evz006.
54. Loman, N. J., & Watson, M. (2015). Successful test launch for nanopore sequencing. *Nature Methods, 12*, 303–304. https://doi.org/10.1038/nmeth.3327.
55. Mikheyev, A. S., & Tin, M. M. (2014). A first look at the Oxford nanopore MinION sequencer. *Molecular Ecology Resources, 14*, 1097–1102. https://doi.org/10.1111/1755-0998.12324.
56. Ayub, M., & Bayley, H. (2012). Individual RNA base recognition in immobilized oligonucleotides using a protein nanopore. *Nano Letters, 12*, 5637–5643. https://doi.org/10.1021/nl3027873.
57. Johnson, S. S., Zaikova, E., Goerlitz, D. S., Bai, Y., & Tighe, S. W. (2017). Real-time DNA sequencing in the Antarctic dry valleys using the Oxford nanopore sequencer. *Journal of Biomolecular Techniques: JBT, 28*, 2–7. https://doi.org/10.7171/jbt.17-2801-009.
58. McIntyre, A. B. R., et al. (2016). Nanopore sequencing in microgravity. *NPJ Microgravity, 2*, 16035. https://doi.org/10.1038/npjmgrav.2016.35.

59. Runtuwene, L. R., Tuda, J. S. B., Mongan, A. E., & Suzuki, Y. (2019). On-site MinION sequencing. *Advances in Experimental Medicine and Biology, 1129*, 143–150. https://doi.org/10.1007/978-981-13-6037-4_10.
60. Walter, M. C., et al. (2017). MinION as part of a biomedical rapidly deployable laboratory. *Journal of Biotechnology, 250*, 16–22. https://doi.org/10.1016/j.jbiotec.2016.12.006.
61. Jain, M., Olsen, H. E., Paten, B., & Akeson, M. (2016). The Oxford nanopore MinION: Delivery of nanopore sequencing to the genomics community. *Genome Biology, 17*, 239. https://doi.org/10.1186/s13059-016-1103-0.
62. Narola, J., Pandey, S. N., Glick, A., & Chen, Y. W. (2013). Conditional expression of TGF-beta1 in skeletal muscles causes endomysial fibrosis and myofibers atrophy. *PLoS One, 8*, e79356. https://doi.org/10.1371/journal.pone.0079356. PONE-D-13-27811 [pii].
63. Cho, D. S., & Doles, J. D. (2017). Single cell transcriptome analysis of muscle satellite cells reveals widespread transcriptional heterogeneity. *Gene, 636*, 54–63. https://doi.org/10.1016/j.gene.2017.09.014.
64. Zeng, W., et al. (2016). Single-nucleus RNA-seq of differentiating human myoblasts reveals the extent of fate heterogeneity. *Nucleic Acids Research, 44*, e158. https://doi.org/10.1093/nar/gkw739.
65. Dell'Orso, S., et al. (2019). Single cell analysis of adult mouse skeletal muscle stem cells in homeostatic and regenerative conditions. *Development, 146*. https://doi.org/10.1242/dev.174177.
66. Winokur, S. T., et al. (2003). Expression profiling of FSHD muscle supports a defect in specific stages of myogenic differentiation. *Human Molecular Genetics, 12*, 2895–2907.
67. Lemmers, R. J., et al. (2010). A unifying genetic model for facioscapulohumeral muscular dystrophy. *Science, 329*, 1650–1653. https://doi.org/10.1126/science.1189044.
68. Snider, L., et al. (2010). Facioscapulohumeral dystrophy: Incomplete suppression of a retrotransposed gene. *PLoS Genetics, 6*, e1001181. https://doi.org/10.1371/journal.pgen.1001181.
69. Jones, T. I., et al. (2015). Individual epigenetic status of the pathogenic D4Z4 macrosatellite correlates with disease in facioscapulohumeral muscular dystrophy. *Clinical Epigenetics, 7*, 37. https://doi.org/10.1186/s13148-015-0072-6.
70. Himeda, C. L., et al. (2014). Myogenic enhancers regulate expression of the facioscapulohumeral muscular dystrophy-associated DUX4 gene. *Molecular and Cellular Biology, 34*, 1942–1955. https://doi.org/10.1128/MCB.00149-14.
71. van den Heuvel, A., et al. (2019). Single-cell RNA sequencing in facioscapulohumeral muscular dystrophy disease etiology and development. *Human Molecular Genetics, 28*, 1064–1075. https://doi.org/10.1093/hmg/ddy400.
72. Ritchie, M. E., et al. (2015). limma powers differential expression analyses for RNA-sequencing and microarray studies. *Nucleic Acids Research, 43*, e47–e47. https://doi.org/10.1093/nar/gkv007.
73. Gautier, L., Cope, L., Bolstad, B. M., & Irizarry, R. A. (2004). affy – Analysis of Affymetrix GeneChip data at the probe level. *Bioinformatics, 20*, 307–315. https://doi.org/10.1093/bioinformatics/btg405.
74. Dunning, M. J., Smith, M. L., Ritchie, M. E., & Tavare, S. (2007). beadarray: R classes and methods for Illumina bead-based data. *Bioinformatics, 23*, 2183–2184. https://doi.org/10.1093/bioinformatics/btm311.
75. Bolstad, B. M., Irizarry, R. A., Astrand, M., & Speed, T. P. (2003). A comparison of normalization methods for high density oligonucleotide array data based on variance and bias. *Bioinformatics, 19*, 185–193. https://doi.org/10.1093/bioinformatics/19.2.185.
76. Carvalho, B. S., & Irizarry, R. A. (2010). A framework for oligonucleotide microarray preprocessing. *Bioinformatics, 26*, 2363–2367. https://doi.org/10.1093/bioinformatics/btq431.
77. Warnes, G. R., Bolker, B., Bonebakker, L., Gentleman, R., Huber, W., Liaw, A., Lumley, T., et al. (2009). *gplots: Various R programming tools for plotting data*. R package version 2.
78. Student. (1908). The probable error of a mean. *Biometrika*.

79. Fisher, R. (1919). A. XV.—The correlation between relatives on the supposition of Mendelian inheritance. *Transactions of the Royal Society of Edinburgh, 52*, 399–433. https://doi.org/10.1017/S0080456800012163.
80. Benjamini, Y., & Hochberg, Y. (1995). Controlling the false discovery rate: A practical and powerful approach to multiple testing. *Journal of the Royal Statistical Society: Series B (Methodological), 57*, 289–300. https://doi.org/10.2307/2346101.
81. Bonferroni, C. (1936). Teoria statistica delle classi e calcolo delle probabilità. *Pubblicazioni del R Istituto Superiore di Scienze Economiche e Commerciali di Firenze, 8*, 3–62.
82. Schadt, E. E., Turner, S., & Kasarskis, A. (2010). A window into third-generation sequencing. *Human Molecular Genetics, 19*, R227–R240. https://doi.org/10.1093/hmg/ddq416.
83. Eisenstein, M. (2012). Oxford nanopore announcement sets sequencing sector abuzz. *Nature Biotechnology, 30*, 295–296. https://doi.org/10.1038/nbt0412-295.
84. Kim, D., et al. (2013). TopHat2: Accurate alignment of transcriptomes in the presence of insertions, deletions and gene fusions. *Genome Biology, 14*, R36. https://doi.org/10.1186/gb-2013-14-4-r36.
85. Trapnell, C., et al. (2012). Differential gene and transcript expression analysis of RNA-seq experiments with TopHat and Cufflinks. *Nature Protocols, 7*, 562–578. https://doi.org/10.1038/nprot.2012.016.
86. Trapnell, C., et al. (2010). Transcript assembly and quantification by RNA-Seq reveals unannotated transcripts and isoform switching during cell differentiation. *Nature Biotechnology, 28*, 511–515. https://doi.org/10.1038/nbt.1621.
87. Langmead, B., & Salzberg, S. L. (2012). Fast gapped-read alignment with Bowtie 2. *Nature Methods, 9*, 357–359. https://doi.org/10.1038/nmeth.1923.
88. Adjeroh, D., Bell, T., & Mukherjee, A. (2008). *The Burrows-Wheeler Transform: Data compression, suffix arrays, and pattern matching*. New York: Springer.
89. Ferragina, P., & Manzini, G. (2001). An experimental study of a compressed index. *Information Sciences, 135*, 13–28. https://doi.org/10.1016/S0020-0255(01)00098-6.
90. Dobin, A., et al. (2013). STAR: Ultrafast universal RNA-seq aligner. *Bioinformatics (Oxford, England), 29*, 15–21. https://doi.org/10.1093/bioinformatics/bts635.
91. Kim, D., Langmead, B., & Salzberg, S. L. (2015). HISAT: A fast spliced aligner with low memory requirements. *Nature Methods, 12*, 357–360. https://doi.org/10.1038/nmeth.3317.
92. Pertea, M., Kim, D., Pertea, G. M., Leek, J. T., & Salzberg, S. L. (2016). Transcript-level expression analysis of RNA-seq experiments with HISAT, StringTie and Ballgown. *Nature Protocols, 11*, 1650–1667. https://doi.org/10.1038/nprot.2016.095.
93. Pertea, M., et al. (2015). StringTie enables improved reconstruction of a transcriptome from RNA-seq reads. *Nature Biotechnology, 33*, 290–295. https://doi.org/10.1038/nbt.3122.
94. Frazee, A. C., et al. (2015). Ballgown bridges the gap between transcriptome assembly and expression analysis. *Nature Biotechnology, 33*, 243–246. https://doi.org/10.1038/nbt.3172.
95. Li, H., et al. (2009). The Sequence Alignment/Map format and SAMtools. *Bioinformatics (Oxford, England), 25*, 2078–2079. https://doi.org/10.1093/bioinformatics/btp352.
96. Wang, L., Wang, S., & Li, W. (2012). RSeQC: Quality control of RNA-seq experiments. *Bioinformatics, 28*, 2184–2185. https://doi.org/10.1093/bioinformatics/bts356.
97. Mortazavi, A., Williams, B. A., McCue, K., Schaeffer, L., & Wold, B. (2008). Mapping and quantifying mammalian transcriptomes by RNA-Seq. *Nature Methods, 5*, 621–628. https://doi.org/10.1038/nmeth.1226.
98. Li, B., Ruotti, V., Stewart, R. M., Thomson, J. A., & Dewey, C. N. (2010). RNA-Seq gene expression estimation with read mapping uncertainty. *Bioinformatics, 26*, 493–500. https://doi.org/10.1093/bioinformatics/btp692.
99. Li, B., & Dewey, C. N. (2011). RSEM: Accurate transcript quantification from RNA-Seq data with or without a reference genome. *BMC Bioinformatics, 12*, 323. https://doi.org/10.1186/1471-2105-12-323.

100. Dempster, A. P., Laird, N. M., & Rubin, D. B. (1976). Maximum likelihood from incomplete data via the EM algorithm. *Journal of the Royal Statistical Society. Series B (Methodological), 39*(1), 1–38.
101. Anders, S., Pyl, P. T., & Huber, W. (2015). HTSeq—A Python framework to work with high-throughput sequencing data. *Bioinformatics (Oxford, England), 31*, 166–169. https://doi.org/10.1093/bioinformatics/btu638.
102. Liao, Y., Smyth, G. K., & Shi, W. (2014). featureCounts: An efficient general purpose program for assigning sequence reads to genomic features. *Bioinformatics, 30*, 923–930. https://doi.org/10.1093/bioinformatics/btt656.
103. Lawrence, M., et al. (2013). Software for computing and annotating genomic ranges. *PLoS Computational Biology, 9*, e1003118. https://doi.org/10.1371/journal.pcbi.1003118.
104. Soneson, C., Love, M. I., & Robinson, M. D. (2015). Differential analyses for RNA-seq: Transcript-level estimates improve gene-level inferences. *F1000Research, 4*, 1521. https://doi.org/10.12688/f1000research.7563.1.
105. Robinson, M. D., McCarthy, D. J., & Smyth, G. K. (2010). edgeR: A bioconductor package for differential expression analysis of digital gene expression data. *Bioinformatics, 26*, 139–140. https://doi.org/10.1093/bioinformatics/btp616.
106. Love, M. I., Huber, W., & Anders, S. (2014). Moderated estimation of fold change and dispersion for RNA-seq data with DESeq2. *Genome Biology, 15*, 550. https://doi.org/10.1186/s13059-014-0550-8.
107. Nelder, J. A., & Wedderburn, R. W. M. (1972). Generalized linear models. *Journal of the Royal Statistical Society. Series A (General), 135*, 370. https://doi.org/10.2307/2344614.
108. Wald, A. (1945). Sequential tests of statistical hypotheses. *The Annals of Mathematical Statistics, 16*, 117–186. https://doi.org/10.1214/aoms/1177731118.
109. Feng, J., et al. (2012). GFOLD: A generalized fold change for ranking differentially expressed genes from RNA-seq data. *Bioinformatics, 28*, 2782–2788. https://doi.org/10.1093/bioinformatics/bts515.
110. Tarazona, S., et al. (2015). Data quality aware analysis of differential expression in RNA-seq with NOISeq R/Bioc package. *Nucleic Acids Research, 43*, e140. https://doi.org/10.1093/nar/gkv711.
111. Eberwine, J., et al. (1992). Analysis of gene expression in single live neurons. *Proceedings of the National Academy of Sciences of the United States of America, 89*, 3010–3014. https://doi.org/10.1073/pnas.89.7.3010.
112. Hwang, B., Lee, J. H., & Bang, D. (2018). Single-cell RNA sequencing technologies and bioinformatics pipelines. *Experimental and Molecular Medicine, 50*, 96. https://doi.org/10.1038/s12276-018-0071-8.
113. Van Der Maaten, L., & Hinton, G. (2008). Visualizing data using t-SNE. *Journal of Machine Learning Research, 9*, 2579–2605.
114. Butler, A., Hoffman, P., Smibert, P., Papalexi, E., & Satija, R. (2018). Integrating single-cell transcriptomic data across different conditions, technologies, and species. *Nature Biotechnology, 36*, 411–420. https://doi.org/10.1038/nbt.4096.
115. Gatto, S., Puri, P. L., & Malecova, B. (2017). Single cell gene expression profiling of skeletal muscle-derived cells. *Methods in Molecular Biology, 1556*, 191–219. https://doi.org/10.1007/978-1-4939-6771-1_10.
116. Banerji, C. R. S., et al. (2017). PAX7 target genes are globally repressed in facioscapulohumeral muscular dystrophy skeletal muscle. *Nature Communications, 8*, 2152. https://doi.org/10.1038/s41467-017-01200-4.
117. Stahl, P. L., et al. (2016). Visualization and analysis of gene expression in tissue sections by spatial transcriptomics. *Science, 353*, 78–82. https://doi.org/10.1126/science.aaf2403.
118. Saletore, Y., et al. (2012). The birth of the Epitranscriptome: Deciphering the function of RNA modifications. *Genome Biology, 13*, 175. https://doi.org/10.1186/gb-2012-13-10-175.

Chapter 6
Approaches to Studying the microRNAome in Skeletal Muscle

Alyson A. Fiorillo and Christopher R. Heier

6.1 Theory and History of the Technique

Muscle is a highly plastic tissue that needs to rapidly undergo dramatic changes in gene expression patterns in order to maintain homeostasis. This requires a delicate balance between satellite cell proliferation, myotube formation and differentiation, and muscle degeneration/regeneration. The disruption of these pathways drives muscle disorders and diseases; this includes dystrophies, inflammatory myopathies, sarcopenia, and cachexia. Thus, identifying factors that regulate muscle gene expression programs is essential to understanding muscle health and function and may uncover new therapeutic targets. Since the discovery of microRNAs (miRNAs) [1–3], it has become well established that they are key regulatory factors which fine-tune gene expression patterns in all cell and tissue types. As we gain new insight into the function of miRNAs, their essential role as posttranscriptional regulatory elements that drive proper muscle function becomes increasingly apparent. As has been observed in the X-linked genetic diseases Duchenne and Becker muscular dystrophies (DMD and BMD, respectively), the chronic dysregulation of miRNAs can exacerbate disease [4–11]. In this chapter we will explore the role of miRNAs in skeletal muscle and the importance of harnessing the power of miRNA profiling to understand how different perturbations to muscle (i.e. exercise, injury, or genetic defects) affect the muscle miRNAome and how the miRNAome, in turn, can yield valuable information about the overall health of muscle.

miRNAs are small noncoding RNAs, approximately 22 nucleotides in length. They primarily function by binding to the 3′ untranslated region (UTR) of a

A. A. Fiorillo (✉) · C. R. Heier
Department of Genomics and Precision Medicine, George Washington University School of Medicine and Health Sciences, Washington, DC, USA

Center for Genetic Medicine Research, Children's National Medical Center, Washington, DC, USA
e-mail: afiorillo@childrensnational.org

J. G. Burniston, Y.-W. Chen (eds.), *Omics Approaches to Understanding Muscle Biology*, Methods in Physiology, https://doi.org/10.1007/978-1-4939-9802-9_6

messenger RNA (mRNA) transcript from a target gene. The binding between a miRNA and the 3′ UTR of a transcript downregulates protein expression from that transcript by either inhibiting protein translation (imperfect complementary binding of the seed sequence) or by promoting mRNA decay (perfect complementary binding of the seed sequence) [12].

In most organisms there are only a small fraction of miRNAs as compared to the number of mRNAs and proteins. In the human genome, it is estimated there are roughly 20,000 protein-coding genes [13], while the number of annotated miRNAs is approximately 2700 (from miRbase V22, http://www.mirbase.org/), although a recent publication suggests this number may actually be as high as 3700 [14]. miRNAs are very stable and highly conserved across species—these factors contribute to their appeal as disease-specific biomarkers [15, 16]. miRNAs are estimated to regulate approximately >60–70% of the mammalian genome [17] and are commonly dysregulated in disease which makes them attractive therapeutic targets [18].

The basic miRNA processing pathway is illustrated in Fig. 6.1 and is reviewed in detail in Refs. [19, 20]. Mature miRNAs are ~22 nucleotides (nt) in length; however, they originate from much longer transcripts. The first step in the miRNA processing pathway is the cleavage of the primary transcript (termed pri-miRNA) by the enzyme

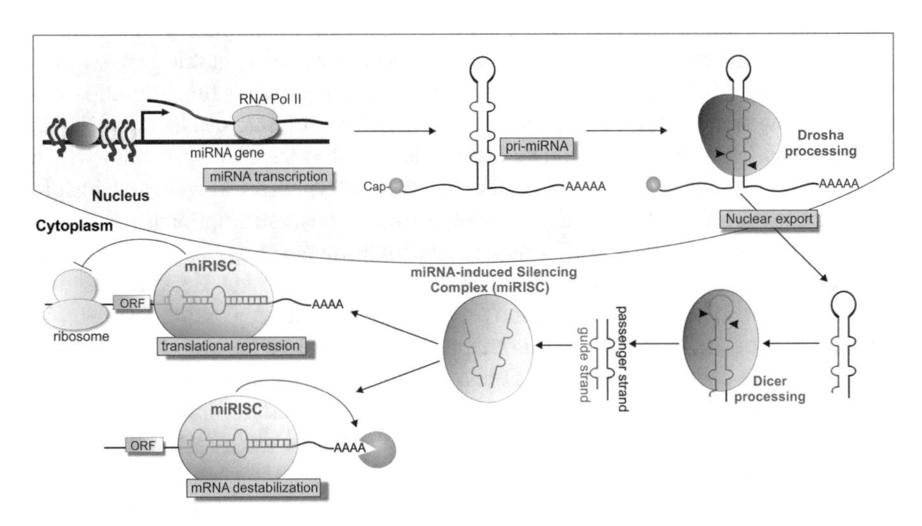

Fig. 6.1 Schematic of miRNA biogenesis. miRNAs are transcribed by RNA polymerase II (RNA Pol II) into a pri-miR, which possesses both a poly(A) tail and a 5′ cap. The enzyme Drosha cleaves the pri-miRNA into a 70–100-nucleotide (nt) hairpin structure termed the pre-miR. The pre-miR translocates into the cytoplasm (via Exportin 5) and is further cleaved by Dicer into two mature miRNAs that are approximately 22 nt. These miRNAs are transferred to the AGO proteins to form the miRNA-induced silencing complex (miRISC). Strand selection occurs, where the dominant strand termed "guide strand" (shown in red throughout the schematic) remains, and the non-active strand or "passenger strand" (shown in black throughout the schematic) is discarded. The RISC complex is then activated and engages the seed sequences of the miRNA with the 3′ UTR of the target mRNA leading to translational repression or mRNA destabilization and decay

Drosha—this yields a pre-miRNA hairpin structure that is approximately 70–100 nt in length. The pre-miR is translocated from the nucleus to the cytoplasm by Exportin 5 [20] and is cleaved by Dicer in combination with trans-activation response RNA-binding protein (TRBP) into two mature miRNAs of approximately 22 nt with short 3′ overhangs [20]. These two mature miRNAs are termed the -3p and -5p strands. Dicer transfers the duplex to one of the four human Argonaute (AGO) proteins, requiring AGO to undergo conformation changes to enable binding of the duplex [21]. Strand selection subsequently occurs where one of the strands (called the passenger strand) is discarded. This allows for activation of the RNA-induced silencing complex (RISC) which contains both AGO and one of the mature miRNAs (called the guide strand or leading strand) and the presentation of the seed sequence of the miRNA (this is most often nucleotides #2–8 of the miRNA). This seed sequence can subsequently interact with its target mRNA, typically within the 3′ UTR of the target transcript. Each miRNA can have hundreds or thousands of targets and thus represents a complex system for "fine-tuning" gene expression patterns [22, 23]. Binding of a miRNA to its target most commonly causes translational repression of that transcript, but sometimes it triggers mRNA decay. The triggering of either translational repression or mRNA decay depends on whether the miRNA sequence is an imperfect match (translational repression) or perfect match (mRNA decay) to the target mRNA sequence [24, 25].

The selection of the guide and passenger strand from the original pre-miRNA can differ between tissues, and the ratio can also change depending on the disease state [26, 27]. It is also important to understand the historical nomenclature to describe the two mature miRNAs produced from the pre-miRNA (i.e., miR∗ vs. miR-3p/-5p). A comprehensive guide for miRNA nomenclature is summarized in a review paper [28].

Described as "micromanagers of gene expression" [29], it is not surprising that miRNAs play a key role in modulating the complex regulatory circuits involved in skeletal muscle formation, maintenance, and physiological and pathological signaling programs [30, 31]. Some miRNAs are ubiquitously expressed, while others are expressed in a tissue-specific manner [32, 33]. miRNAs that are either exclusively expressed or enriched in striated muscle are referred to as myomiRs [34]. The myomiR family includes eight miRNAs: miR-1, miR-133a, miR-133b, miR-206, miR-208a, miR-208b, miR-486, and miR-499 [35–39]. myomiRs are expressed in both skeletal and cardiac muscle with two exceptions: miR-208a is expressed only in cardiac muscle, while miR-206 is exclusively expressed in skeletal muscle. Here, we will focus on skeletal muscle myomiRs.

Several myomiRs are encoded by DNA within introns of muscle-specific genes, which facilitate their expression specifically within muscle. For example, muscle-specific miR-208b and miR-499 are encoded within introns of the myosin heavy chain genes *MYH7* and *MYH7B*, respectively [38, 39], while miR-486 is considered muscle-enriched and is located within an intron of ankyrin 1 (*ANK1*) [37]. The other myomiRs are organized into bicistronic clusters and transcribed together [40]. For example, miR-1-2 and miR-133a-1 are encoded within an intron of the mindbomb E3 ubiquitin ligase 1 *(MIB1)* gene, while miR-1-1 and miR-133a-2 are found in the intron of a muscle-enriched gene encoded by open reading frame 166 on chromosome 20 (C20orf166). In contrast, a cluster consisting of miR-206 and miR-133b resides in an intergenic or non-protein coding space on chromosome 6. Regardless of their genomic locus, the transcriptional regulation of all of myomiRs is known to be driven by myogenic regulatory factors (MRFs) such as MYOD, MYOG, MYF5, MYF6, and SRF [41, 42].

It is important to note that a subset of miRNAs, whose expression is not restricted to muscle, also play key roles in muscle signaling and development. These miRNAs include miR-26a [43], miR-27b [44], miR-29 [45], miR-125b [46], miR-155 [47], miR-128a [48], miR-181 [49], miR-24 [50], miR-378 [51], miR221/222 [52], and miR-214 [53]. In a disease state, the balance of miRNAs is dysregulated: miRNAs not normally found in the muscle are increased [7–9], while miRNAs that are critical to maintaining muscle homeostasis are decreased [35]. This has been demonstrated in Duchenne muscular dystrophy (DMD) where both myomiRs and non-myomiRs are highly dysregulated. Studies have shown regeneration-specific miRNAs are elevated (miR-31, miR-34c, miR-206, miR-335, miR-449, and miR-494) [6, 10] and degenerative miRNAs are downregulated (miR-1, miR-29c, and miR-135a) [10] which is linked to an increase in fibrosis. Further, several studies have found that inflammatory miRNAs are elevated in DMD including miR-222, miR-223, miR-146a, miR-146b, miR-382, miR-320a, miR-142-5p, 142-3p, miR-301a, miR-324-3p, miR-455-5p, miR-455-3p, miR-497, and miR-652 [6, 8–10]. Deregulation of miRNA expression is a common feature of several skeletal muscle disorders [7], and specific treatment regimens (e.g., anti-inflammatories, exon skipping) bring some of these dysregulated miRNAs back toward homeostasis [6, 8, 9]. A detailed view of how miRNAs affect normal and diseased muscle is summarized in Fig. 6.2.

6.2 Major Applications

6.2.1 miRNA Profiling

One advantage to profiling miRNAs is that they are much more stable than mRNAs and can be recovered from formalin-fixed paraffin-embedded sections or other sources that typically show low overall RNA quality. Consistent sample processing and RNA extraction methods are critically important to the quality of results from miRNA profiling, however. With that in mind, samples should be handled with care to ensure consistent results and to avoid possible miRNA degradation [54]. In the following sections, we will discuss the major methods of miRNA isolation and profiling, as well as the advantages and disadvantages of each method.

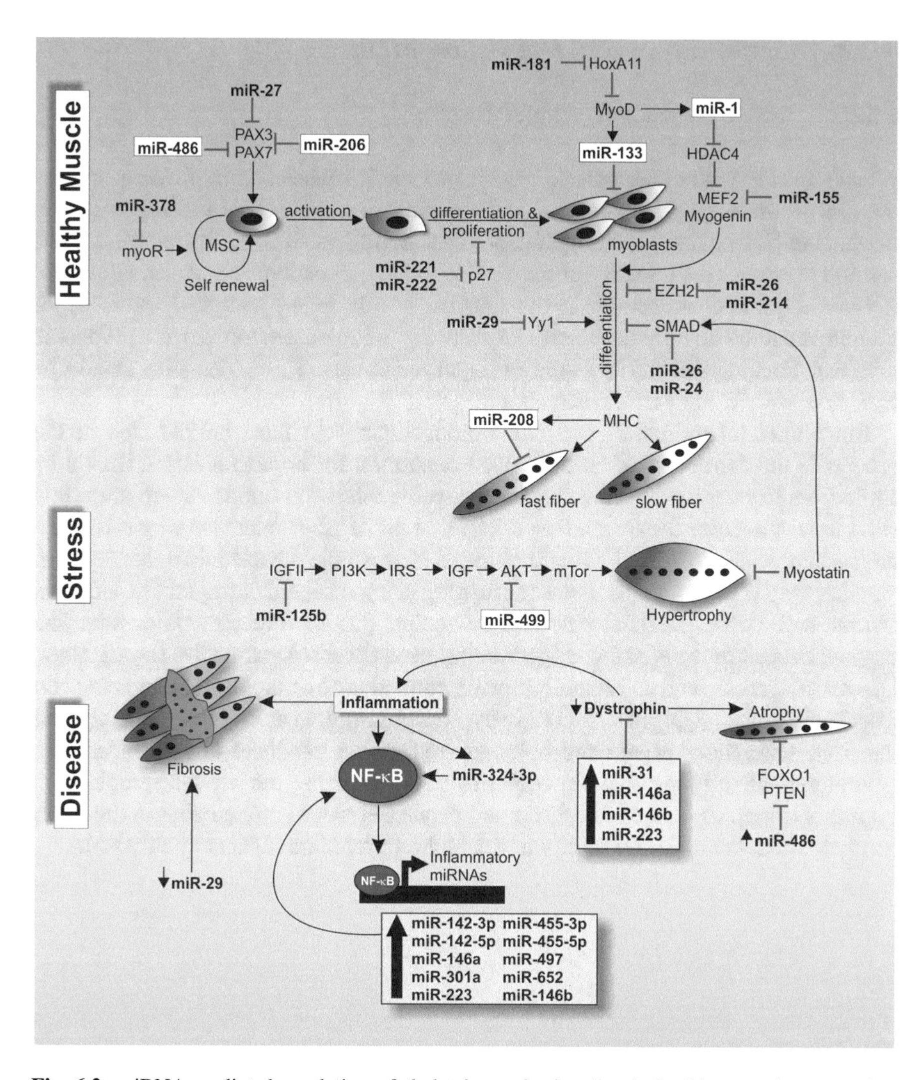

Fig. 6.2 miRNA-mediated regulation of skeletal muscle signaling in healthy muscle, stressed or exercised muscle, and disease. Schematic represents gene regulation by miRNAs known to be involved in muscle development and muscle growth (hypertrophy). The bottom two panels illustrate how dysregulated miRNAs in disease feed into different pathways (inflammation, atrophy, fibrosis) to exacerbate disease. Red bars represent inhibition. Canonical myomiRs are shown in white boxes. Adapted from Refs. [9, 30]

6.2.2 *Sample Types and Extraction Methods*

6.2.2.1 Muscle Biopsies or Whole Muscles

When working with either muscle biopsies or whole muscles from different species (i.e., rat, mouse, dog), we recommend freezing muscle in liquid nitrogen-cooled isopentane (also named 2-methylbutane) and storing at −80 °C for later use. This method preserves the integrity of the muscle tissue for sectioning and also allows for RNA and protein extraction. If the muscle is only being analyzed for RNA or protein, it can be directly flash frozen by placing the dissected muscle in a cryogenic tube and freezing directly in liquid nitrogen for at least 1 min. Samples should be stored at −80 °C until RNA extraction.

Since skeletal muscle is a highly fibrous tissue, the most critical step in the process is the disruption of all the cells. Preparation for homogenization should be carried out on dry ice or under liquid nitrogen-cooled conditions. There are a few ways to homogenize the frozen tissue. The first, most labor-intensive way is to use a mortar and pestle cooled with liquid nitrogen to crush the muscle into a fine powder. Another way is to first perform tissue crushing using a Liquid Nitrogen Cooled Mini Mortar and Pestle Set (Bel-Art) and following this to directly homogenize the crushed tissues in the specific reagent being used for RNA extraction (i.e., TRIzol, QIAzol reagents) with a TissueRuptor or equivalent handheld homogenizer with disposable probes (Qiagen). Alternatively, frozen samples can be placed directly into the RNA extraction reagent and homogenized with a handheld homogenizer. Our laboratories have found that the combination of crushing the muscle sample with Liquid Nitrogen Cooled Mini Mortar and Pestle Set and homogenizing in the RNA extraction reagent (TRIzol) results in the highest yield and highest quality RNA.

6.2.2.2 Muscle Cell Lines

If using muscle cell lines for miRNA profiling, the procedure is much easier and less labor intensive. Prior to RNA extraction, media and cell debris should be rinsed off of cells with PBS and then cells can either be frozen for later use (−80 °C) or RNA extraction can be immediately performed by pipetting the RNA extraction reagent (i.e., TRIzol) directly onto cells in plates or dishes. Differentiated myotubes grown in six-well plates typically provide a concentration of 200–500 ng/ul of RNA (when resuspended in 25–30 μL of RNase-free water), which is more than enough to work with for profiling experiments.

6.2.2.3 Additional Considerations

Before beginning RNA extraction, it is important to apply RNase inhibitor spray to the laboratory bench and to utilize RNase-free water, tubes, and equipment to prevent RNA degradation. For RNA extraction, the most common reagents and kits utilize chemical extraction with concentrated chaotropic salts such as guanidine thiocyanate (TRIzol and QIAzol). As an optional step, there are commercial kits available to enrich for miRNAs (miRNeasy, Qiagen; *mir*Vana, Ambion; and PureLink, Thermo Fisher Scientific). These kits add a solid-phase extraction procedure on silica columns. However, if you have limited samples and plan on analyzing both miRNA and mRNA, we recommend skipping these miRNA enrichment steps since the columns will only bind small RNAs and mRNAs will be lost in the wash steps. To increase the yield of miRNA recovered using these methods, we recommend performing the isopropanol precipitation step at −20 °C overnight or over the weekend before performing the final ethanol wash and RNA re-suspension.

Unlike mRNA profiling, formalin-fixed paraffin embedded (FFPE) tissue can be utilized for miRNA profiling [55]. mRNA becomes fragmented in FFPE samples; however miRNA is quite stable, largely because it is less subject to RNase-mediated degradation [56]. This is a clear advantage of utilizing miRNA profiling in a clinical sample cohort where oftentimes FFPE samples are the only sample type available. The procedure for recovering miRNAs from this sample type begins with cutting FFPE samples into thick sections (~20 μm). Depending on the width of the cross sections, as little as one section (i.e., for a muscle sample taken at autopsy) and as many as four 20 μm sections (for a small muscle biopsy), or up to 35 mg of unsectioned muscle, can be used. Following deparaffinization and protease treatment [55, 57], miRNA can be recovered from these tissues by commercially available kits such as RecoverAll Total Nucleic Acid Isolation Kit (Thermo Fisher Scientific), High Pure miRNA Isolation Kit (Roche), or miRNeasy FFPE Kit (Qiagen).

6.2.3 Sample Heterogeneity

Muscle tissues contain a heterogeneous population of cell types [58]. Depending on the experimental setup and question, it may be necessary to purify a specific cell type from muscle tissue (i.e., satellite cells, macrophages, differentiated muscle) or purify a specific area from a muscle section (i.e., damaged area, high inflammation area). This can be achieved by using fluorescence-activated cell sorting (FACS), myofiber isolation, or laser capture microdissection (LCM) [59, 60]. In FACS a specific cell type is selected from a mixture of live cells using a fluorescently labeled antibody against a specific cell marker, most often a receptor, and miRNA profiling is subsequently performed [60, 61]. Another way to use this method is to purify genetically labeled cells from a transgenic mouse [62]. Alternatively, LCM can be used on histological sections of dissected muscle to cut out a specific area of interest and purify RNA from these regions. This may be particularly useful if your specimen has variable pathology, such as in studies on the effects of local injury (e.g., notexin injection) or in assessing different stages of muscle development.

6.2.4 Concentration and Quality Assessment

Once RNA is extracted, it is important to check quality and concentration. As a general rule, each milligram of tissue should yield about 1 μg of total RNA. All profiling methods discussed here can use total RNA, so it is not absolutely necessary to assess miRNA quantity separately. The most simplistic way to determine overall RNA quality and quantity is using spectrophotometry (i.e., NanoDrop). RNA has a maximum absorbance at 260 nm, and this reading will yield the RNA concentration. Historically, the ratio of absorbance at 260 to the absorbance at 280 nm has been used as a measure of purity, with a ratio of 2.0 denoting a pure RNA sample. For profiling experiments, RNA that has a purity of 1.7 or higher can generally be used. Once RNA concentrations are determined, it is common practice to dilute all samples to the same concentration. For a more in-depth analysis of RNA quality, the Bioanalyzer 2100 (Agilent) can be used with the small RNA chip. This chip can selectively quantify miRNAs in absolute amounts [pg/μL] and as a relative percentage of small RNA [%]. However, the estimation of miRNA abundance by this method may only be accurate when overall RNA integrity is very high [63]. It is also possible to assess miRNA extraction efficiency by incorporating a "spike-in" synthetic miRNA from another species (most commonly *C. elegans*) at an early step in the RNA isolation [64]. Quantification of these spike-in miRNAs in the RNA recovered from muscle tissues/cell lines can serve as an RNA extraction efficiency control, though other normalization methods are preferred for downstream miRNA quantification.

6.2.5 miRNA Profiling Methods

There are three well-defined approaches for miRNA profiling: quantitative reverse transcription PCR (qRT-PCR), hybridization-based methods, and RNA sequencing. Deciding which method to choose will need to be based on your experimental goals and your available resources. Figure 6.3 summarizes the basic miRNA profiling platforms.

6.2.6 qRT-PCR-Based Methods

Low-density quantitative reverse transcriptase polymerase chain reaction (qRT-PCR) arrays have emerged as a first-line method of choice by many for miRNA profiling. This is because of the reproducibility of their data in subsequent single-gene qPCR experiments, their increased sensitivity (dynamic range of seven log10 units), and their ability to utilize a small amount of input RNA (as little as 25 picograms). Whereas other platforms require much greater starting material [65],

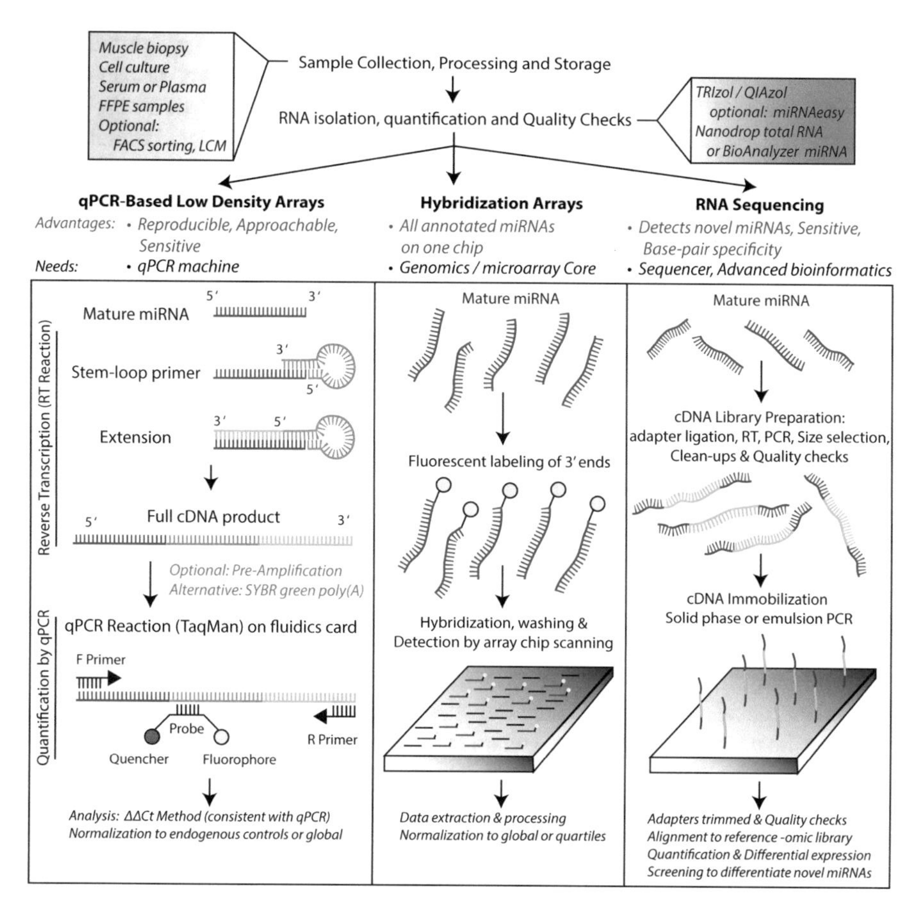

Fig. 6.3 miRNA profiling platforms. Schematic summarizes the overall methodologies for miRNA profiling of muscle or related samples. Each of the columns presents the concepts and processes involved in one of three main profiling methods: qPCR-based low-density arrays (stem-loop RT method depicted here), hybridization arrays, and RNA sequencing. The qPCR-based methods are very approachable for those with prior experience in single-gene qPCR. Concepts behind the stem-loop primer RT method are depicted here, an approach that is used by low-density microfluidics cards to assay approximately 750 miRNAs per set. Next, hybridization arrays are widely available through genomics cores and can simultaneously assay hundreds of pre-annotated miRNAs from multiple species on a single miRNA gene array chip. Finally, RNA sequencing provides a highly precise method with the ability to detect novel miRNAs or highly similar miRNAs at single base-pair specificity

it has been demonstrated that this method can detect miRNA content at the single-cell level [66].

As miRNAs are too small to utilize standard methods to perform qRT-PCR, the reverse transcription reaction has been modified to generate complementary DNA (cDNA) that can be subsequently used in qRT-PCR. There are two main methods of priming. Chen et al. developed a technique using stem-loop reverse transcription

(RT) priming to synthesize cDNA from miRNA, followed by TaqMan real-time PCR analysis [67]. In the reverse transcription reaction, the stem-loop primer serves to prime the cDNA reaction as well as extend the size of the mature miRNA: it consists of a constant region that forms a stem loop and a variable six-nucleotide extension. The stem-loop portion of the primer extends the 18–22-nucleotide miRNA to >60 nucleotides to allow for traditional PCR in subsequent steps. The six-nucleotide extension is the reverse complement of the last six nucleotides on the 3′ end of the miRNA of interest and provides specificity. A second method to convert mature miRNA to cDNA involves the addition of a poly(A) tail using *E. coli* poly(A) polymerase and then performing qPCR using a miRNA-specific forward primer and a universal poly(T) adaptor reverse primer [68].

To perform large-scale miRNA profiling using these methods, the RT-PCR primers are pooled together and thus convert all target miRNAs into cDNA. This is followed by performing qRT-PCR with pre-plated PCR primers that are distributed across multi-well plates or microfluidic cards containing nanoliter-scale wells. There is also an option for a pre-amplification step, adding around 10–14 cycles of amplification before samples are run on qPCR array plates. Examples of commercially available qRT-PCR arrays include TaqMan low-density miRNA array cards (stem-loop priming, Thermo Fisher Scientific) and miScript miRNA PCR Arrays (poly(A) priming, Qiagen).

Data analysis for qRT-PCR arrays is performed using a method commonly known as the ΔΔCt method, enabling the comparison of groups by using one group as the normalizer [69]. This calculation assumes that the amplification product in qPCR doubles after every PCR cycle. The abundance, then, is inversely related to the cycle (Ct) number in which you detect the signal from your amplicon. In basic terms, the ΔΔCt is calculated (1) from normalizing the cycle in which your target amplifies to the cycle in which your housekeeping reference RNA amplified (called ΔCt) and (2) from then normalizing your "treatment" sample to your "control" sample (called ΔΔCt). More specifically, since each cycle of PCR represents a doubling of your miRNA target, this is all actually done in logarithmic space where the value at each cycle (Ct) is actually 2^Ct. The ΔCt is equal to Ct_{miRNA} minus $Ct_{Housekeeping}$, and ΔΔCt is equal to ΔCt_{Test} minus $\Delta Ct_{Control}$. To get a final, relative quantification, the ΔΔCt result is converted back into a linear space by the eq. $2^{\wedge(-\Delta\Delta Ct)}$. Different platforms have different software for performing this relative quantification in parallel and in an automated fashion. It is imperative, however, that amplification curves for each individual miRNA assay are checked before formalizing your results and/or using global normalization methods, as these programs don't always flag improper amplification in the "noise" range. Not performing the proper quality checks can therefore lead to the false appearance that the miRNA expression in one group is significantly different, when in reality the Ct values from the target miRNAs or from miRNAs included as normalization controls are simply background noise. This process of quality checks can add labor to the profiling analysis; however it can also increase confidence and insight by allowing one to directly view the amplification curve data for each individual result.

In terms of data normalization, there are a few strategies that can be used: (1) normalization to stably expressed endogenous reference small RNAs, (2) global normalization to the average of all Ct values from array experiments, and (3) normalization to an external spike-in synthetic oligonucleotide. The first strategy, normalization to reference small RNAs, commonly uses small nuclear/nucleolar RNAs such as snoRNAs or RNU6A and 6B. For best results, the geometric mean of multiple control genes should be used in order to minimize effects of any variation or noise in the individual reference RNAs. As housekeeping RNAs can still be affected by certain diseases or conditions [70], this is always a good idea for qPCR-based gene expression analyses and is often included as an automated option in expression analysis software. If normalization/control RNAs are suspected to be changing or are not well-defined in a sample type, several statistical programs exist that can analyze profiling data within an experiment and rank candidate reference genes in order of stability. These include BestKeeper [71], GeNorm [72], and NormFinder [73]. These programs also take into account intra- and intergroup variability. As an alternative to using specific endogenous controls, another common method is to use global normalization where the mean expression value of all miRNAs is used as the normalization factor. Care should be taken to quality check each individual miRNA using this approach. This global normalization method has been reported to reduce technical variation and provide an accurate appreciation of biological change [74]. A third normalization method is to normalize to a spike-in synthetic oligonucleotide. This method is frequently used as a processing control to measure miRNA extraction efficiency, and in some instances, it may be used for gene expression when overall miRNA content is expected to be quite different in control vs. treatment samples [75]. However it is less favorable than the other methods for normalizing gene expression, as a spike-in method only corrects for either extraction efficiency or reverse transcription efficiency and not for expression variability within the original biological sample.

6.2.7 Hybridization-Based Methods

6.2.7.1 miRNA Microarrays

Microarrays were the first technology to be utilized for high-throughput analysis of miRNA expression. Their advantages are the number of samples that can be analyzed at once, that one array can cover more miRNAs from more species, and that many institutions already have microarray core facilities to run them. miRNA microarrays can screen hundreds of target sequences within a single sample, including both precursor and mature miRNAs. The design of miRNA arrays is different from that of other spotted microarrays, however, as the differing characteristics of miRNAs must be taken into account. First, mature miRNAs are not polyadenylated, meaning that a label (biotin, fluorescence) cannot be incorporated into miRNAs via oligo (dT)-primed reverse transcription. Second, there is minimal sequence available on miRNAs to enable hybridization since miRNAs are typically only 18–22 nt in length. This means probe design must also be restricted to this length. The restriction in probe length is challenging because, similar to PCR, designing such short hybridization probes leads to variable melting temperature (T_m) caused by highly variable GC richness (i.e., 70% GC-rich vs. 20% GC-rich miRNAs require very different hybridization temperatures) meaning that conditions cannot be optimized for all miRNAs being detected by the array.

Carlo M. Croce and colleagues described the first miRNA profiling array [76] which measured miRNA precursors rather than mature miRNAs. This method utilizes biotin labeling of miRNAs via RT-PCR, priming the reaction with oligo d (T) primers to turn poly(A) tailed transcripts into biotinylated cDNA as miRNA precursors are polyadenylated similar to protein coding transcripts. In this method labeled miRNAs are hybridized to a slide that is spotted with 40 nt oligo probes specific for all human and murine pre-miRNAs. Hybridized miRNAs are detected after adding a streptavidin-conjugated Alexa Fluor 647 label using a hybridization chip reader.

A handful of the techniques to perform miRNA arrays involve an enzyme-mediated labeling of miRNAs. The first described mature miRNA array method utilizes a T4 RNA ligase to couple the 3′ end of miRNAs to a fluorescent (Cy3)-modified dinucleotide, while the 5′ end hybridizes to capture probes that are spotted onto slides [77]. A reference oligonucleotide for all mature miRNAs labeled with Cy5 is included for normalization. This method requires a dephosphorylation step to remove the 5′ phosphate from miRNAs before hybridization; without this step miRNAs are at risk for circularization. Another method of probe labeling is poly (A) extension of the 3′ end using poly(A) polymerase [78], following which a fluorophore-conjugated oligonucleotide is ligated to the miRNA using a splinted or bridged ligation technique [79, 80]. Here, ligation of an oligo to the 3′ end of a miRNA is facilitated by a second oligonucleotide (the "bridge oligo") that hybridizes to both the 3′ end of the miRNA and the 5′ end of the "capture" oligonucleotide. This technique avoids the issue of circularization; however, poly(A) tailing is generally variable in how many A's are added to the miRNA, which could create variable hybridization properties between miRNAs.

Another method is to modify the miRNAs after they are hybridized to their respective probes. Nelson and colleagues developed a miRNA array platform using RNA-primed, array-based, Klenow enzyme (RAKE) to label miRNAs post-hybridization [81]. Here, DNA capture probes bound to slides have a spacer sequence with three thymidine bases adjacent to the miRNA-binding region. After miRNA hybridization, the microarray is treated with DNA exonuclease I, which degrades the probes not bound by miRNAs. The Klenow fragment of DNA polymerase I, an enzyme that can use RNA to prime its activity, is added with biotinylated dATP, which becomes incorporated complementary to the three thymidines in the capture probe. The amount of capture miRNA is subsequently determined by the addition of a fluorescently conjugated streptavidin. Similar approaches have been adapted by others [82, 83].

Like PCR, variable melting temperature (T_m) of the GC-rich content of miRNAs is an obstacle for miRNA hybridization probes. To help reduce T_m and make hybridization of miRNAs to capture probes more consistent, substituting locked nucleic acid (LNA)-modified probes in place of DNA probes has proven to be useful [84] as these synthetic RNA/DNA molecules increase the overall thermostability of oligonucleotide probes.

Normalization of miRNA hybridization arrays can be performed in a variety of ways as reviewed by Pradervand et al. [40]. One common normalization method is to use global miRNA expression. This applies the assumption that specific miRNAs in a sample may vary, but the overall miRNA content stays the same [85, 86]. This method has been well validated and utilized in the literature. Other methods for data normalization are described by Hua et al. [87] and include quantile normalization or variance stabilizing normalization (VSN). The goal of the quantile method is to make the distribution of probe intensities for each array in a set of arrays the same. VSN assumes that most miRNAs are not differentially expressed. Both of these normalization methods are available as part of the open-source Bioconductor bioinformatics software (https://www.bioconductor.org/). It is recommended, however, that you discuss your normalization method with your bioinformatics team to choose the best normalization method for your dataset.

6.2.7.2 NanoString nCounter Gene Expression System

A more recent development in hybridization-based miRNA profiling technology is the NanoString nCounter, which applies a unique color-coded "fingerprint" to miRNA-specific probes [88]. This technology utilizes a biotin-labeled 3′ capture probe and a 5′ reporter probe with a color code that is unique to each miRNA. The result is the formation of a tripartite structure consisting of a miRNA in between two bound probes. Unbound probes are removed via affinity purification, and bound complexes are immobilized onto a streptavidin-coated slide, followed by imaging and counting of bound reporters. Data analysis can be performed using nSolver™, a software provided by NanoString Technologies. This technique has high sensitivity and high specificity, avoids amplification bias, and has the advantage of being able to discriminate between similar miRNA variants.

6.2.8 miRNA Sequencing

The development of next-generation sequencing platforms provides us with powerful but complex methods to profile miRNA expression. Major strengths of these platforms include their ability to detect novel miRNAs and their ability to precisely identify miRNAs down to single base-pair differences. This can provide the ability to differentiate between similar miRNAs within the same "family" and between isomiR's which possess slight modifications of the reference miRNA sequence.

A general pipeline for an RNA-seq project runs as follows: sample production, RNA isolation, RNA sub-type selection, library preparation, quality checks, next-generation sequencing, data quality checks, alignment of sequencing data to the genome, and data visualization and analysis. Initially, RNA isolation and quantification can be performed similarly to messenger RNA experiments, as high-quality miRNA can be easily isolated from muscle samples using total RNA isolation methods and reagents such as TRIzol. Subsequently, miRNA isolation kits such as miRNeasy (Qiagen) are commonly used in order to both enrich for miRNA and further purify the samples for subsequent enzymatic reactions. Library preparation is performed after miRNA isolation and consists of ligating linker sequences to the miRNAs, followed by reverse transcription to convert these into cDNAs, then PCR amplification of these cDNAs, and subsequent purification through gel-based size selection. Once the library preparation is complete, the miRNA-seq reaction can be performed using any of several next-generation sequencing platforms. Examples of these platforms include high-throughput sequencers such as the HiSeq 2000 (Illumina) and SOLiD (ABI), as well as smaller-scale sequencers such as Ion Torrent (Invitrogen) and MiSeq (Illumina).

Note, the miRNA-seq workflow becomes much more complex than qPCR-based methods due to the added steps of library preparation and the subsequent computational bioinformatics steps needed to analyze and interpret the data. Some tool suites have been developed to help streamline the data analysis process. These include publicly available, open-access suites such as the Galaxy Mississippi tool suite (https://mississippi.snv.jussieu.fr/), as well as platform-specific tool suites provided by vendors. A brief description of the steps and general concepts follows. First, compressed sequencing data files may need to be converted to file types used for analyses (.fastq file types). Adapter or linker sequences which were ligated to the miRNAs during library preparation then need to be "trimmed" from the sequencing data to isolate the miRNA sequences alone and the data needs to be quality checked; these can be performed using software tools such as miRge [89] and Cutadapt [90]. Once data clears quality checks, it needs to be aligned to a reference genome or RNA library so that the miRNAs can be properly recognized, positioned, and subsequently quantified or visualized. This alignment is performed using a genome alignment tool such as Bowtie [91] or sRbowtie. Quantification can then be performed to detect differences in the expression levels of miRNAs, using software such as DESeq2 [92] or DEXUS [93]. In order to identify novel miRNAs within a tissue or sample, further tools such as miRDeep2 [94], miRanalyzer [95], or novoMiRank [96] are needed. The purpose of these tools is to reduce false-positive identification of other small RNA species as miRNAs, and they can integrate information such as related pre-miRNA sequences or predicted secondary structures.

During RNA isolation and selection, different methods can be used to select for specific subsets or general types of small RNA species. For example, a method known as CLIP-seq uses biochemical techniques to immunoprecipitate miRNA species bound by specific proteins such as Argonaute (AGO), a protein component of the RNA-induced silencing complex (RISC). Integrating biochemical techniques such as this has the benefit of removing some non-miRNA small RNA species and of providing additional biological insight; however it also removes unbound miRNAs from the sample. Alternatively, strictly size-based approaches to enrich samples for miRNAs can be used; however additional small RNAs are then included in the sequencing analysis which can complicate the data analysis. Computational approaches need to be subsequently applied to help minimize incorrect identification of such small RNAs as miRNAs.

Future developments in next-generation sequencing will enhance the abilities of miRNA-seq, improve data storage, and make the technologies involved more broadly accessible. This provides another advantage for next-generation sequencing approaches—whereas qPCR- and hybridization-based technologies are largely static, the technologies and analysis capabilities of sequencing are constantly evolving. Moving forward, new technologies in single-molecule sequencing (SMS) are being developed with the potential to improve both speed and data bias in comparison with current platforms [97].

6.2.9 miRNA Databases

Once data is generated from profiling, there are a few tools that will help in data interpretation. The most utilized and comprehensive general database is miRBase (www.miRbase.org, v22) [98, 99]. miRBase provides information on the predicted hairpin portion of the miRNA transcript (pre-miR), on its genomic locus, and on the mature miRNA sequence. It provides comprehensive information for all annotated miRNAs, including nomenclature, sequence data, predicted gene targets, and validated targets of each miRNA. There are also other tools that help with miRNA target prediction, with mining the literature for validated targets, and with determining the functional significance of miRNAs identified in your profiling data. Some useful databases include the following: (1) TargetScan [100], DIANA-TarBase [101], miRTarBase [102, 103], and miRDB [104, 105] which provide target prediction, (2) miRWalk 7.0 [106, 107] which provides validated miRNA-target interactions and miRNA binding sites, and (3) DIANA miRPath [108] and miR2Disease [109] which integrate miRNA profiling data with relevant pathways and diseases. We also recommend using the UCSC Genome Browser to look at potential transcription factors that regulate the miRNAs from your profiling data, as this can give further clues into the pathways affected by experimental conditions [110]. A comprehensive review of these databases is available [111, 112].

6.3 Limitations

In addition to their benefits, each of the miRNA profiling methods described here has its own limitations. For qRT-PCR-based methods, these limitations stem from the requirement of a predefined, separate, and specific primer/probe combination to be used in an enzymatic reaction for each unique miRNA. One consequence of this is that novel miRNAs cannot be identified using this method. Additionally, since each miRNA requires its own qPCR assay, there is a limit to how many miRNAs can be assayed at once. Typical qRT-PCR array card systems can assay up to around 384 or 768 miRNAs at once. As a consequence, fewer miRNAs are assayed at once, species-specific array cards are needed, and miRNA precursors are not typically included.

The limitations of hybridization-based miRNA microarray methods include the following: they require more starting material than qRT-PCR-based methods, they have lower sensitivity and specificity than other methods making it difficult to discern miRNAs with similar sequences (e.g., miR-146a-5p vs. miR-146b-5p), and they do not enable the absolute quantification of miRNA abundance. Rather, using this approach is an effective way to more broadly compare "healthy" to "diseased" muscle or non-exercised to exercised muscle. A further limitation of the NanoString nCounter gene expression system is that it is quite expensive compared to other methods and the instrumentation required to run it is less broadly available.

Major disadvantages of miRNA sequencing platforms center around their cost and the advanced computational bioinformatics resources needed to properly analyze the large datasets produced. Additionally, upstream biases are inherently introduced to the datasets through library preparation methods which preferentially select for specific types of RNAs. Post-acquisition, storage of the very large datasets and files produced by next-generation sequencing represents a substantial problem for the field. Advances in computing storage, in data compression, and in minimizing or modifying the type of data files needed for long-term memory storage will all help to address this issue. From the data analysis side, improvements to software and algorithms are frequently being made at a rapid pace. Because of this, available programs can quickly become out of date. Before planning and executing a next-generation sequencing project, it is thus important to consult with a computational bioinformatics expert and the recent literature to plan your project according to the most trusted analysis workflows and software.

6.4 Vital Future Directions

As miRNAs are a relatively new class of biomolecule, they represent an exciting area of research that we are only in the early stages of understanding. As several of the technologies we have described are already well established due to redundancy with mRNA techniques, here we will focus on one emerging technique and several exciting applications that are rapidly evolving. Digital qPCR will be discussed as a newer technique which enables sensitive and absolute, as opposed to relative, quantification of miRNAs. Next, serum miRNA biomarkers have the potential to improve diagnosis, improve translation of preclinical studies, and improve decisions on clinical drug approvals. Finally, as a newer class of biomolecule that is dysregulated in disease, miRNAs provide us with an exciting new class of therapeutic targets.

The concept of digital PCR was first developed in 1992 [113] (then termed PCR with limiting dilution) and later defined as digital PCR [114] when utilized to quantify cancer-causing mutations. While this technology was usurped by the development of quantitative real-time PCR, it is now reemerging due to the recent development of better instruments and chemistry which have made it a simpler and more practical technique. Digital PCR, in contrast to real-time PCR, provides absolute quantification of the target miRNA based on partitioning of individual molecules into thousands of replicate reactions at low dilution. This results in zero or one target miRNA in most reactions. At the end of the PCR reaction, the absolute concentration of the miRNA (in copies/μl) is determined by Poisson statistical analysis of the number of positive and negative reactions. The main advantages of digital PCR are as follows: (1) the ability to determine absolute quantification without external references [115]; (2) no requirement for an endogenous control for normalization, and (3) a high degree of sensitivity and precision as compared to qPCR. Because of its high sensitivity, digital PCR may enable researchers to accurately determine absolute physiological concentration of a given miRNA within serum or other biofluids, where traditional intracellular housekeeping genes are not present for relative normalization. As we learn more about the muscle miRNAome, research is shifting toward the use of serum biomarkers as surrogate measures for molecular changes in skeletal muscle. Through this evolution, digital PCR may become an increasingly powerful tool to detect low copy number miRNAs that can accurately predict and reflect the overall condition of skeletal muscle as it relates to disease, exercise, aging, and overall health.

A particularly exciting area of miRNA research is the development of serum miRNAs as minimally invasive biomarkers, which has the potential to improve the diagnosis, monitoring, and treatment of muscle diseases. Loosely defined, a biomarker is a measurement that reflects a biological activity such as a disease or a drug response. The discovery that miRNAs are present and highly stable in patient serum provides a new class of highly quantifiable molecule which can be objectively and routinely measured [16]. Whereas serum proteins are relatively unstable and degrade during freeze-thaw cycles, serum can be repeatedly freeze-thawed at least six times or even exposed to RNases with no discernable effect on miRNA levels. In muscle diseases, miRNAs are released by damaged muscle into serum in a manner that may help to predict disease state or pharmacodynamic response. For instance, myomiRs are muscle-specific miRNAs and are released from damaged muscle. Accordingly, myomiRs are detected at significantly increased levels in serum from muscular dystrophy patients versus healthy volunteers and can further differentiate milder BMD patients from more severe DMD patients [116]. In addition to myomiRs, pathology-specific miRNAs are upregulated in certain disease states and may provide mechanism-defined biomarkers. For example, inflammatory miRNAs such as miR-146a are increased in dystrophic muscle of both BMD/DMD patients and animal models, and patient serum levels of these same miRNAs are found to predict a response to anti-inflammatory drugs in other diseases [8, 9, 117]. Moving forward, a major goal is to develop serum miRNAs into objective and minimally invasive surrogate outcome measures in clinical drug trials. Essentially, this will require serum miRNAs to be validated and a biological rationale built which enables them to be used as substitutes for clinically meaningful endpoints which predict the effectiveness of therapies.

In addition to monitoring disease, miRNAs provide us with exciting new strategies to treat disease. Two basic strategies for miRNA-based therapeutics involve either the addition of a beneficial miRNA exogenously to improve pathology or the inhibition of a pathological miRNA. The first of these is a bit more straightforward. If a decrease in a particular miRNA is found to drive disease pathology, for example, by resulting in upregulation of a pathological protein normally repressed by that miRNA, then replacing that miRNA with a "miRNA mimic" may improve pathology. The second strategy typically involves developing an antisense oligo to inhibit a miRNA whose presence or upregulation is found to promote pathology. A strength of this strategy is that, since it is targeting an RNA-based molecule, highly specific drugs can be designed that target the specific nucleic acid sequences of that miRNA. Toward this end, a diverse number of strategies and chemistries are being pursued to ultimately produce a new generation of miRNA-based therapeutics. These include antagomiRs which are antisense oligos complementary to the miRNA [18], optimized "Tough Decoy" inhibitors [118, 119], miRNA "sponges" which present decoy miRNA binding sites [120], and "target protectors" or "masks" which bind 3′ UTRs to block miRNA binding without disrupting expression of the target [121]. In recent years, at least nine miRNA-based therapeutics have entered preclinical or clinical development, as reviewed in [122]. Moving forward, it will be interesting to see how these early trials and next-generation drugs perform as we uncover the full potential of miRNA-based therapeutics.

References

1. Lee, R. C., Feinbaum, R. L., & Ambros, V. (1993). The *C. elegans* heterochronic gene lin-4 encodes small RNAs with antisense complementarity to lin-14. *Cell, 75*, 843–854.
2. Reinhart, B. J., Slack, F. J., Basson, M., Pasquinelli, A. E., Bettinger, J. C., Rougvie, A. E., Horvitz, H. R., & Ruvkun, G. (2000). The 21-nucleotide let-7 RNA regulates developmental timing in *Caenorhabditis elegans*. *Nature, 403*, 901–906.
3. Wightman, B., Ha, I., & Ruvkun, G. (1993). Posttranscriptional regulation of the heterochronic gene lin-14 by lin-4 mediates temporal pattern formation in *C. elegans*. *Cell, 75*, 855–862.
4. Cacchiarelli, D., Incitti, T., Martone, J., Cesana, M., Cazzella, V., Santini, T., Sthandier, O., & Bozzoni, I. (2011). miR-31 modulates dystrophin expression: New implications for Duchenne muscular dystrophy therapy. *EMBO Reports, 12*, 136–141.
5. Cacchiarelli, D., Legnini, I., Martone, J., Cazzella, V., D'Amico, A., Bertini, E., & Bozzoni, I. (2011). miRNAs as serum biomarkers for Duchenne muscular dystrophy. *EMBO Molecular Medicine, 3*, 258–265.
6. Cacchiarelli, D., Martone, J., Girardi, E., Cesana, M., Incitti, T., Morlando, M., Nicoletti, C., Santini, T., Sthandier, O., Barberi, L., Auricchio, A., Musaro, A., & Bozzoni, I. (2010).

MicroRNAs involved in molecular circuitries relevant for the Duchenne muscular dystrophy pathogenesis are controlled by the dystrophin/nNOS pathway. *Cell Metabolism, 12*, 341–351.
7. Eisenberg, I., Eran, A., Nishino, I., Moggio, M., Lamperti, C., Amato, A. A., Lidov, H. G., Kang, P. B., North, K. N., Mitrani-Rosenbaum, S., Flanigan, K. M., Neely, L. A., Whitney, D., Beggs, A. H., Kohane, I. S., & Kunkel, L. M. (2007). Distinctive patterns of microRNA expression in primary muscular disorders. *Proceedings of the National Academy of Sciences of the United States of America, 104*, 17016–17021.
8. Fiorillo, A. A., Heier, C. R., Novak, J. S., Tully, C. B., Brown, K. J., Uaesoontrachoon, K., Vila, M. C., Ngheim, P. P., Bello, L., Kornegay, J. N., Angelini, C., Partridge, T. A., Nagaraju, K., & Hoffman, E. P. (2015). TNF-alpha-Induced microRNAs control dystrophin expression in Becker muscular dystrophy. *Cell Reports, 12*, 1678–1690.
9. Fiorillo, A. A., Tully, C. B., Damsker, J. M., Nagaraju, K., Hoffman, E. P., & Heier, C. R. (2018). Muscle miRNAome shows suppression of chronic inflammatory miRNAs with both prednisone and vamorolone. *Physiological Genomics, 50*, 735–745.
10. Greco, S., De Simone, M., Colussi, C., Zaccagnini, G., Fasanaro, P., Pescatori, M., Cardani, R., Perbellini, R., Isaia, E., Sale, P., Meola, G., Capogrossi, M. C., Gaetano, C., & Martelli, F. (2009). Common micro-RNA signature in skeletal muscle damage and regeneration induced by Duchenne muscular dystrophy and acute ischemia. *FASEB Journal: Official Publication of the Federation of American Societies for Experimental Biology, 23*, 3335–3346.
11. Liu, N., Williams, A. H., Maxeiner, J. M., Bezprozvannaya, S., Shelton, J. M., Richardson, J. A., Bassel-Duby, R., & Olson, E. N. (2012). microRNA-206 promotes skeletal muscle regeneration and delays progression of Duchenne muscular dystrophy in mice. *The Journal of Clinical Investigation, 122*, 2054–2065.
12. Valencia-Sanchez, M. A., Liu, J., Hannon, G. J., & Parker, R. (2006). Control of translation and mRNA degradation by miRNAs and siRNAs. *Genes and Development, 20*, 515–524.
13. Ezkurdia, I., Juan, D., Rodriguez, J. M., Frankish, A., Diekhans, M., Harrow, J., Vazquez, J., Valencia, A., & Tress, M. L. (2014). Multiple evidence strands suggest that there may be as few as 19,000 human protein-coding genes. *Human Molecular Genetics, 23*, 5866–5878.
14. Londin, E., Loher, P., Telonis, A. G., Quann, K., Clark, P., Jing, Y., Hatzimichael, E., Kirino, Y., Honda, S., Lally, M., Ramratnam, B., Comstock, C. E., Knudsen, K. E., Gomella, L., Spaeth, G. L., Hark, L., Katz, L. J., Witkiewicz, A., Rostami, A., Jimenez, S. A., Hollingsworth, M. A., Yeh, J. J., Shaw, C. A., SE, M. K., Bray, P., Nelson, P. T., Zupo, S., Van Roosbroeck, K., Keating, M. J., Calin, G. A., Yeo, C., Jimbo, M., Cozzitorto, J., Brody, J. R., Delgrosso, K., Mattick, J. S., Fortina, P., & Rigoutsos, I. (2015). Analysis of 13 cell types reveals evidence for the expression of numerous novel primate- and tissue-specific microRNAs. *Proceedings of the National Academy of Sciences of the United States of America, 112*, E1106–E1115.
15. Lewis, B. P., Burge, C. B., & Bartel, D. P. (2005). Conserved seed pairing, often flanked by adenosines, indicates that thousands of human genes are microRNA targets. *Cell, 120*, 15–20.
16. Mitchell, P. S., Parkin, R. K., Kroh, E. M., Fritz, B. R., Wyman, S. K., Pogosova-Agadjanyan, E. L., Peterson, A., Noteboom, J., O'Briant, K. C., Allen, A., Lin, D. W., Urban, N., Drescher, C. W., Knudsen, B. S., Stirewalt, D. L., Gentleman, R., Vessella, R. L., Nelson, P. S., Martin, D. B., & Tewari, M. (2008). Circulating microRNAs as stable blood-based markers for cancer detection. *Proceedings of the National Academy of Sciences of the United States of America, 105*, 10513–10518.
17. Friedman, R. C., Farh, K. K., Burge, C. B., & Bartel, D. P. (2009). Most mammalian mRNAs are conserved targets of microRNAs. *Genome Research, 19*, 92–105.
18. Krutzfeldt, J., Rajewsky, N., Braich, R., Rajeev, K. G., Tuschl, T., Manoharan, M., & Stoffel, M. (2005). Silencing of microRNAs in vivo with 'antagomirs'. *Nature, 438*, 685–689.
19. Davis-Dusenbery, B. N., & Hata, A. (2010). Mechanisms of control of microRNA biogenesis. *Journal of Biochemistry, 148*, 381–392.

20. Kim, V. N., Han, J., & Siomi, M. C. (2009). Biogenesis of small RNAs in animals. *Nature Reviews. Molecular Cell Biology, 10*, 126–139.
21. Wang, Y., Juranek, S., Li, H., Sheng, G., Wardle, G. S., Tuschl, T., & Patel, D. J. (2009). Nucleation, propagation and cleavage of target RNAs in Ago silencing complexes. *Nature, 461*, 754–761.
22. Chi, S. W., Zang, J. B., Mele, A., & Darnell, R. B. (2009). Argonaute HITS-CLIP decodes microRNA-mRNA interaction maps. *Nature, 460*, 479–486.
23. Hafner, M., Landthaler, M., Burger, L., Khorshid, M., Hausser, J., Berninger, P., Rothballer, A., Ascano, M., Jr., Jungkamp, A. C., Munschauer, M., Ulrich, A., Wardle, G. S., Dewell, S., Zavolan, M., & Tuschl, T. (2010). Transcriptome-wide identification of RNA-binding protein and microRNA target sites by PAR-CLIP. *Cell, 141*, 129–141.
24. Djuranovic, S., Nahvi, A., & Green, R. (2012). miRNA-mediated gene silencing by translational repression followed by mRNA deadenylation and decay. *Science, 336*, 237–240.
25. Meijer, H. A., Kong, Y. W., Lu, W. T., Wilczynska, A., Spriggs, R. V., Robinson, S. W., Godfrey, J. D., Willis, A. E., & Bushell, M. (2013). Translational repression and eIF4A2 activity are critical for microRNA-mediated gene regulation. *Science, 340*, 82–85.
26. Bang, C., Batkai, S., Dangwal, S., Gupta, S. K., Foinquinos, A., Holzmann, A., Just, A., Remke, J., Zimmer, K., Zeug, A., Ponimaskin, E., Schmiedl, A., Yin, X., Mayr, M., Halder, R., Fischer, A., Engelhardt, S., Wei, Y., Schober, A., Fiedler, J., & Thum, T. (2014). Cardiac fibroblast-derived microRNA passenger strand-enriched exosomes mediate cardiomyocyte hypertrophy. *The Journal of Clinical Investigation, 124*, 2136–2146.
27. Meijer, H. A., Smith, E. M., & Bushell, M. (2014). Regulation of miRNA strand selection: Follow the leader? *Biochemical Society Transactions, 42*, 1135–1140.
28. Pritchard, C. C., Cheng, H. H., & Tewari, M. (2012). MicroRNA profiling: Approaches and considerations. *Nature Reviews Genetics, 13*, 358–369.
29. Bartel, D. P., & Chen, C. Z. (2004). Micromanagers of gene expression: The potentially widespread influence of metazoan microRNAs. *Nature Reviews Genetics, 5*, 396–400.
30. Ballarino, M., Morlando, M., Fatica, A., & Bozzoni, I. (2016). Non-coding RNAs in muscle differentiation and musculoskeletal disease. *The Journal of Clinical Investigation, 126*, 2021–2030.
31. Guller, I., & Russell, A. P. (2010). MicroRNAs in skeletal muscle: Their role and regulation in development, disease and function. *The Journal of Physiology, 588*, 4075–4087.
32. Lagos-Quintana, M., Rauhut, R., Yalcin, A., Meyer, J., Lendeckel, W., & Tuschl, T. (2002). Identification of tissue-specific microRNAs from mouse. *Current Biology: CB, 12*, 735–739.
33. Lee, R. C., & Ambros, V. (2001). An extensive class of small RNAs in *Caenorhabditis elegans*. *Science, 294*, 862–864.
34. McCarthy, J. J. (2008). MicroRNA-206: The skeletal muscle-specific myomiR. *Biochimica et Biophysica Acta, 1779*, 682–691.
35. McCarthy, J. J., & Esser, K. A. (2007). MicroRNA-1 and microRNA-133a expression are decreased during skeletal muscle hypertrophy. *Journal of Applied Physiology (1985), 102*, 306–313.
36. Sempere, L. F., Freemantle, S., Pitha-Rowe, I., Moss, E., Dmitrovsky, E., & Ambros, V. (2004). Expression profiling of mammalian microRNAs uncovers a subset of brain-expressed microRNAs with possible roles in murine and human neuronal differentiation. *Genome Biology, 5*, R13.
37. Small, E. M., O'Rourke, J. R., Moresi, V., Sutherland, L. B., McAnally, J., Gerard, R. D., Richardson, J. A., & Olson, E. N. (2010). Regulation of PI3-kinase/Akt signaling by muscle-enriched microRNA-486. *Proceedings of the National Academy of Sciences of the United States of America, 107*, 4218–4223.
38. van Rooij, E., Quiat, D., Johnson, B. A., Sutherland, L. B., Qi, X., Richardson, J. A., Kelm, R. J., Jr., & Olson, E. N. (2009). A family of microRNAs encoded by myosin genes governs myosin expression and muscle performance. *Developmental Cell, 17*, 662–673.

39. van Rooij, E., Sutherland, L. B., Qi, X., Richardson, J. A., Hill, J., & Olson, E. N. (2007). Control of stress-dependent cardiac growth and gene expression by a microRNA. *Science, 316*, 575–579.
40. Pradervand, S., Weber, J., Thomas, J., Bueno, M., Wirapati, P., Lefort, K., Dotto, G. P., & Harshman, K. (2009). Impact of normalization on miRNA microarray expression profiling. *RNA, 15*, 493–501.
41. Rao, P. K., Kumar, R. M., Farkhondeh, M., Baskerville, S., & Lodish, H. F. (2006). Myogenic factors that regulate expression of muscle-specific microRNAs. *Proceedings of the National Academy of Sciences of the United States of America, 103*, 8721–8726.
42. Rosenberg, M. I., Georges, S. A., Asawachaicharn, A., Analau, E., & Tapscott, S. J. (2006). MyoD inhibits Fstl1 and Utrn expression by inducing transcription of miR-206. *The Journal of Cell Biology, 175*, 77–85.
43. Dey, B. K., Gagan, J., Yan, Z., & Dutta, A. (2012). miR-26a is required for skeletal muscle differentiation and regeneration in mice. *Genes and Development, 26*, 2180–2191.
44. Crist, C. G., Montarras, D., Pallafacchina, G., Rocancourt, D., Cumano, A., Conway, S. J., & Buckingham, M. (2009). Muscle stem cell behavior is modified by microRNA-27 regulation of Pax3 expression. *Proceedings of the National Academy of Sciences of the United States of America, 106*, 13383–13387.
45. Wei, W., He, H. B., Zhang, W. Y., Zhang, H. X., Bai, J. B., Liu, H. Z., Cao, J. H., Chang, K. C., Li, X. Y., & Zhao, S. H. (2013). miR-29 targets Akt3 to reduce proliferation and facilitate differentiation of myoblasts in skeletal muscle development. *Cell Death and Disease, 4*, e668.
46. Ge, Y., Sun, Y., & Chen, J. (2011). IGF-II is regulated by microRNA-125b in skeletal myogenesis. *The Journal of Cell Biology, 192*, 69–81.
47. Seok, H. Y., Tatsuguchi, M., Callis, T. E., He, A., Pu, W. T., & Wang, D. Z. (2011). miR-155 inhibits expression of the MEF2A protein to repress skeletal muscle differentiation. *The Journal of Biological Chemistry, 286*, 35339–35346.
48. Motohashi, N., Alexander, M. S., Shimizu-Motohashi, Y., Myers, J. A., Kawahara, G., & Kunkel, L. M. (2013). Regulation of IRS1/Akt insulin signaling by microRNA-128a during myogenesis. *Journal of Cell Science, 126*, 2678–2691.
49. Naguibneva, I., Ameyar-Zazoua, M., Polesskaya, A., Ait-Si-Ali, S., Groisman, R., Souidi, M., Cuvellier, S., & Harel-Bellan, A. (2006). The microRNA miR-181 targets the homeobox protein Hox-A11 during mammalian myoblast differentiation. *Nature Cell Biology, 8*, 278–284.
50. Sun, Q., Zhang, Y., Yang, G., Chen, X., Cao, G., Wang, J., Sun, Y., Zhang, P., Fan, M., Shao, N., & Yang, X. (2008). Transforming growth factor-beta-regulated miR-24 promotes skeletal muscle differentiation. *Nucleic Acids Research, 36*, 2690–2699.
51. Gagan, J., Dey, B. K., Layer, R., Yan, Z., & Dutta, A. (2011). MicroRNA-378 targets the myogenic repressor MyoR during myoblast differentiation. *The Journal of Biological Chemistry, 286*, 19431–19438.
52. Cardinali, B., Castellani, L., Fasanaro, P., Basso, A., Alema, S., Martelli, F., & Falcone, G. (2009). Microrna-221 and microrna-222 modulate differentiation and maturation of skeletal muscle cells. *PLoS One, 4*, e7607.
53. Shi, K., Lu, J., Zhao, Y., Wang, L., Li, J., Qi, B., Li, H., & Ma, C. (2013). MicroRNA-214 suppresses osteogenic differentiation of C2C12 myoblast cells by targeting Osterix. *Bone, 55*, 487–494.
54. Ibberson, D., Benes, V., Muckenthaler, M. U., & Castoldi, M. (2009). RNA degradation compromises the reliability of microRNA expression profiling. *BMC Biotechnology, 9*, 102.
55. Doleshal, M., Magotra, A. A., Choudhury, B., Cannon, B. D., Labourier, E., & Szafranska, A. E. (2008). Evaluation and validation of total RNA extraction methods for microRNA expression analyses in formalin-fixed, paraffin-embedded tissues. *The Journal of molecular diagnostics: JMD, 10*, 203–211.

56. Aryani, A., & Denecke, B. (2015). In vitro application of ribonucleases: Comparison of the effects on mRNA and miRNA stability. *BMC Research Notes, 8*, 164.
57. Xi, Y., Nakajima, G., Gavin, E., Morris, C. G., Kudo, K., Hayashi, K., & Ju, J. (2007). Systematic analysis of microRNA expression of RNA extracted from fresh frozen and formalin-fixed paraffin-embedded samples. *RNA, 13*, 1668–1674.
58. Yablonka-Reuveni, Z., & Nameroff, M. (1987). Skeletal muscle cell populations. Separation and partial characterization of fibroblast-like cells from embryonic tissue using density centrifugation. *Histochemistry, 87*, 27–38.
59. Gautam, V., & Sarkar, A. K. (2015). Laser assisted microdissection, an efficient technique to understand tissue specific gene expression patterns and functional genomics in plants. *Molecular Biotechnology, 57*, 299–308.
60. Iyer-Pascuzzi, A. S., & Benfey, P. N. (2010). Fluorescence-activated cell sorting in plant developmental biology. *Methods in Molecular Biology, 655*, 313–319.
61. Coll, M., El Taghdouini, A., Perea, L., Mannaerts, I., Vila-Casadesus, M., Blaya, D., Rodrigo-Torres, D., Affo, S., Morales-Ibanez, O., Graupera, I., Lozano, J. J., Najimi, M., Sokal, E., Lambrecht, J., Gines, P., van Grunsven, L. A., & Sancho-Bru, P. (2015). Integrative miRNA and gene expression profiling analysis of human quiescent hepatic stellate cells. *Scientific Reports, 5*, 11549.
62. Lobo, M. K., Karsten, S. L., Gray, M., Geschwind, D. H., & Yang, X. W. (2006). FACS-array profiling of striatal projection neuron subtypes in juvenile and adult mouse brains. *Nature Neuroscience, 9*, 443–452.
63. Pritchard, C. C., Kroh, E., Wood, B., Arroyo, J. D., Dougherty, K. J., Miyaji, M. M., Tait, J. F., & Tewari, M. (2012). Blood cell origin of circulating microRNAs: A cautionary note for cancer biomarker studies. *Cancer Prevention Research (Philadelphia, Pa.), 5*, 492–497.
64. Kroh, E. M., Parkin, R. K., Mitchell, P. S., & Tewari, M. (2010). Analysis of circulating microRNA biomarkers in plasma and serum using quantitative reverse transcription-PCR (qRT-PCR). *Methods, 50*, 298–301.
65. Chen, Y., Gelfond, J. A., McManus, L. M., & Shireman, P. K. (2009). Reproducibility of quantitative RT-PCR array in miRNA expression profiling and comparison with microarray analysis. *BMC Genomics, 10*, 407.
66. Mestdagh, P., Feys, T., Bernard, N., Guenther, S., Chen, C., Speleman, F., & Vandesompele, J. (2008). High-throughput stem-loop RT-qPCR miRNA expression profiling using minute amounts of input RNA. *Nucleic Acids Research, 36*, e143.
67. Chen, C., Ridzon, D. A., Broomer, A. J., Zhou, Z., Lee, D. H., Nguyen, J. T., Barbisin, M., Xu, N. L., Mahuvakar, V. R., Andersen, M. R., Lao, K. Q., Livak, K. J., & Guegler, K. J. (2005). Real-time quantification of microRNAs by stem-loop RT-PCR. *Nucleic Acids Research, 33*, e179.
68. Shi, R., Sun, Y. H., Zhang, X. H., & Chiang, V. L. (2012). Poly(T) adaptor RT-PCR. *Methods in Molecular Biology, 822*, 53–66.
69. Livak, K. J., & Schmittgen, T. D. (2001). Analysis of relative gene expression data using real-time quantitative PCR and the 2(-Delta Delta C(T)) Method. *Methods, 25*, 402–408.
70. Gee, H. E., Buffa, F. M., Camps, C., Ramachandran, A., Leek, R., Taylor, M., Patil, M., Sheldon, H., Betts, G., Homer, J., West, C., Ragoussis, J., & Harris, A. L. (2011). The small-nucleolar RNAs commonly used for microRNA normalisation correlate with tumour pathology and prognosis. *British Journal of Cancer, 104*, 1168–1177.
71. Pfaffl, M. W., Tichopad, A., Prgomet, C., & Neuvians, T. P. (2004). Determination of stable housekeeping genes, differentially regulated target genes and sample integrity: BestKeeper – Excel-based tool using pair-wise correlations. *Biotechnology Letters, 26*, 509–515.
72. Vandesompele, J., De Preter, K., Pattyn, F., Poppe, B., Van Roy, N., De Paepe, A., & Speleman, F. (2002). Accurate normalization of real-time quantitative RT-PCR data by geometric averaging of multiple internal control genes. *Genome Biology, 3*. https://doi.org/10.1186/gb-2002-3-7-research0034.

73. Andersen, C. L., Jensen, J. L., & Orntoft, T. F. (2004). Normalization of real-time quantitative reverse transcription-PCR data: A model-based variance estimation approach to identify genes suited for normalization, applied to bladder and colon cancer data sets. *Cancer Research, 64*, 5245–5250.
74. Mestdagh, P., Van Vlierberghe, P., De Weer, A., Muth, D., Westermann, F., Speleman, F., & Vandesompele, J. (2009). A novel and universal method for microRNA RT-qPCR data normalization. *Genome Biology, 10*, R64.
75. Roberts, T. C., Coenen-Stass, A. M., & Wood, M. J. (2014). Assessment of RT-qPCR normalization strategies for accurate quantification of extracellular microRNAs in murine serum. *PLoS One, 9*, e89237.
76. Liu, C. G., Calin, G. A., Meloon, B., Gamliel, N., Sevignani, C., Ferracin, M., Dumitru, C. D., Shimizu, M., Zupo, S., Dono, M., Alder, H., Bullrich, F., Negrini, M., & Croce, C. M. (2004). An oligonucleotide microchip for genome-wide microRNA profiling in human and mouse tissues. *Proceedings of the National Academy of Sciences of the United States of America, 101*, 9740–9744.
77. Thomson, J. M., Parker, J., Perou, C. M., & Hammond, S. M. (2004). A custom microarray platform for analysis of microRNA gene expression. *Nature Methods, 1*, 47–53.
78. Goff, L. A., Yang, M., Bowers, J., Getts, R. C., Padgett, R. W., & Hart, R. P. (2005). Rational probe optimization and enhanced detection strategy for microRNAs using microarrays. *RNA Biology, 2*, 93–100.
79. Git, A., Dvinge, H., Salmon-Divon, M., Osborne, M., Kutter, C., Hadfield, J., Bertone, P., & Caldas, C. (2010). Systematic comparison of microarray profiling, real-time PCR, and next-generation sequencing technologies for measuring differential microRNA expression. *RNA, 16*, 991–1006.
80. Maroney, P. A., Chamnongpol, S., Souret, F., & Nilsen, T. W. (2008). Direct detection of small RNAs using splinted ligation. *Nature Protocols, 3*, 279–287.
81. Nelson, P. T., Baldwin, D. A., Scearce, L. M., Oberholtzer, J. C., Tobias, J. W., & Mourelatos, Z. (2004). Microarray-based, high-throughput gene expression profiling of microRNAs. *Nature Methods, 1*, 155–161.
82. Berezikov, E., van Tetering, G., Verheul, M., van de Belt, J., van Laake, L., Vos, J., Verloop, R., van de Wetering, M., Guryev, V., Takada, S., van Zonneveld, A. J., Mano, H., Plasterk, R., & Cuppen, E. (2006). Many novel mammalian microRNA candidates identified by extensive cloning and RAKE analysis. *Genome Research, 16*, 1289–1298.
83. Yeung, M. L., Bennasser, Y., Myers, T. G., Jiang, G., Benkirane, M., & Jeang, K. T. (2005). Changes in microRNA expression profiles in HIV-1-transfected human cells. *Retrovirology, 2*, 81.
84. Castoldi, M., Schmidt, S., Benes, V., Noerholm, M., Kulozik, A. E., Hentze, M. W., & Muckenthaler, M. U. (2006). A sensitive array for microRNA expression profiling (miChip) based on locked nucleic acids (LNA). *RNA, 12*, 913–920.
85. Bissels, U., Wild, S., Tomiuk, S., Holste, A., Hafner, M., Tuschl, T., & Bosio, A. (2009). Absolute quantification of microRNAs by using a universal reference. *RNA, 15*, 2375–2384.
86. Risso, D., Massa, M. S., Chiogna, M., & Romualdi, C. (2009). A modified LOESS normalization applied to microRNA arrays: A comparative evaluation. *Bioinformatics, 25*, 2685–2691.
87. Hua, Y. J., Tu, K., Tang, Z. Y., Li, Y. X., & Xiao, H. S. (2008). Comparison of normalization methods with microRNA microarray. *Genomics, 92*, 122–128.
88. Geiss, G. K., Bumgarner, R. E., Birditt, B., Dahl, T., Dowidar, N., Dunaway, D. L., Fell, H. P., Ferree, S., George, R. D., Grogan, T., James, J. J., Maysuria, M., Mitton, J. D., Oliveri, P., Osborn, J. L., Peng, T., Ratcliffe, A. L., Webster, P. J., Davidson, E. H., Hood, L., & Dimitrov, K. (2008). Direct multiplexed measurement of gene expression with color-coded probe pairs. *Nature Biotechnology, 26*, 317–325.

89. Baras, A. S., Mitchell, C. J., Myers, J. R., Gupta, S., Weng, L. C., Ashton, J. M., Cornish, T. C., Pandey, A., & Halushka, M. K. (2015). miRge – A multiplexed method of processing small RNA-seq data to determine microRNA entropy. *PLoS One, 10*, e0143066.
90. Chen, C., Khaleel, S. S., Huang, H., & Wu, C. H. (2014). Software for pre-processing Illumina next-generation sequencing short read sequences. *Source Code for Biology and Medicine, 9*, 8.
91. Langmead, B. (2010). Aligning short sequencing reads with Bowtie. *Current Protocols in Bioinformatics, Chapter 11*, Unit 11.7.
92. Love, M. I., Huber, W., & Anders, S. (2014). Moderated estimation of fold change and dispersion for RNA-seq data with DESeq2. *Genome Biology, 15*, 550.
93. Klambauer, G., Unterthiner, T., & Hochreiter, S. (2013). DEXUS: Identifying differential expression in RNA-Seq studies with unknown conditions. *Nucleic Acids Research, 41*, e198.
94. Mackowiak, S. D. (2011). Identification of novel and known miRNAs in deep-sequencing data with miRDeep2. *Current Protocols in Bioinformatics Chapter 12*, Unit 12.10.
95. Hackenberg, M., Sturm, M., Langenberger, D., Falcon-Perez, J. M., & Aransay, A. M. (2009). miRanalyzer: A microRNA detection and analysis tool for next-generation sequencing experiments. *Nucleic Acids Research, 37*, W68–W76.
96. Backes, C., Meder, B., Hart, M., Ludwig, N., Leidinger, P., Vogel, B., Galata, V., Roth, P., Menegatti, J., Grasser, F., Ruprecht, K., Kahraman, M., Grossmann, T., Haas, J., Meese, E., & Keller, A. (2016). Prioritizing and selecting likely novel miRNAs from NGS data. *Nucleic Acids Research, 44*, e53.
97. Kapranov, P., Ozsolak, F., & Milos, P. M. (2012). Profiling of short RNAs using Helicos single-molecule sequencing. *Methods in Molecular Biology, 822*, 219–232.
98. Griffiths-Jones, S., Grocock, R. J., van Dongen, S., Bateman, A., & Enright, A. J. (2006). miRBase: microRNA sequences, targets and gene nomenclature. *Nucleic Acids Research, 34*, D140–D144.
99. Griffiths-Jones, S., Saini, H. K., van Dongen, S., & Enright, A. J. (2008). miRBase: Tools for microRNA genomics. *Nucleic Acids Research, 36*, D154–D158.
100. Agarwal, V., Bell, G. W., Nam, J. W., & Bartel, D. P. (2015). Predicting effective microRNA target sites in mammalian mRNAs. *eLife, 4*. https://doi.org/10.7554/eLife.05005.
101. Vlachos, I. S., Paraskevopoulou, M. D., Karagkouni, D., Georgakilas, G., Vergoulis, T., Kanellos, I., Anastasopoulos, I. L., Maniou, S., Karathanou, K., Kalfakakou, D., Fevgas, A., Dalamagas, T., & Hatzigeorgiou, A. G. (2015). DIANA-TarBase v7.0: Indexing more than half a million experimentally supported miRNA:mRNA interactions. *Nucleic Acids Research, 43*, D153–D159.
102. Chou, C. H., Shrestha, S., Yang, C. D., Chang, N. W., Lin, Y. L., Liao, K. W., Huang, W. C., Sun, T. H., Tu, S. J., Lee, W. H., Chiew, M. Y., Tai, C. S., Wei, T. Y., Tsai, T. R., Huang, H. T., Wang, C. Y., Wu, H. Y., Ho, S. Y., Chen, P. R., Chuang, C. H., Hsieh, P. J., Wu, Y. S., Chen, W. L., Li, M. J., Wu, Y. C., Huang, X. Y., Ng, F. L., Buddhakosai, W., Huang, P. C., Lan, K. C., Huang, C. Y., Weng, S. L., Cheng, Y. N., Liang, C., Hsu, W. L., & Huang, H. D. (2018). miRTarBase update 2018: A resource for experimentally validated microRNA-target interactions. *Nucleic Acids Research, 46*, D296–D302.
103. Hsu, S. D., Lin, F. M., Wu, W. Y., Liang, C., Huang, W. C., Chan, W. L., Tsai, W. T., Chen, G. Z., Lee, C. J., Chiu, C. M., Chien, C. H., Wu, M. C., Huang, C. Y., Tsou, A. P., & Huang, H. D. (2011). miRTarBase: A database curates experimentally validated microRNA-target interactions. *Nucleic Acids Research, 39*, D163–D169.
104. Wang, X. (2008). miRDB: A microRNA target prediction and functional annotation database with a wiki interface. *RNA, 14*, 1012–1017.
105. Wong, N., & Wang, X. (2015). miRDB: An online resource for microRNA target prediction and functional annotations. *Nucleic Acids Research, 43*, D146–D152.
106. Dweep, H., & Gretz, N. (2015). miRWalk2.0: A comprehensive atlas of microRNA-target interactions. *Nature Methods, 12*, 697.
107. Parveen, A., Gretz, N., & Dweep, H. (2016). Obtaining miRNA-Target Interaction Information from miRWalk2.0. *Current Protocols in Bioinformatics, 55*, 12.15.11–12.15.27.

108. Vlachos, I. S., Zagganas, K., Paraskevopoulou, M. D., Georgakilas, G., Karagkouni, D., Vergoulis, T., Dalamagas, T., & Hatzigeorgiou, A. G. (2015). DIANA-miRPath v3.0: Deciphering microRNA function with experimental support. *Nucleic Acids Research, 43*, W460–W466.
109. Jiang, Q., Wang, Y., Hao, Y., Juan, L., Teng, M., Zhang, X., Li, M., Wang, G., & Liu, Y. (2009). miR2Disease: A manually curated database for microRNA deregulation in human disease. *Nucleic Acids Research, 37*, D98–D104.
110. Kent, W. J., Sugnet, C. W., Furey, T. S., Roskin, K. M., Pringle, T. H., Zahler, A. M., & Haussler, D. (2002). The human genome browser at UCSC. *Genome Research, 12*, 996–1006.
111. Riffo-Campos, A. L., Riquelme, I., & Brebi-Mieville, P. (2016). Tools for sequence-based miRNA target prediction: What to choose? *International Journal of Molecular Sciences, 17*. https://doi.org/10.3390/ijms17121987.
112. Vlachos, I. S., & Hatzigeorgiou, A. G. (2013). Online resources for miRNA analysis. *Clinical Biochemistry, 46*, 879–900.
113. Sykes, P. J., Neoh, S. H., Brisco, M. J., Hughes, E., Condon, J., & Morley, A. A. (1992). Quantitation of targets for PCR by use of limiting dilution. *BioTechniques, 13*, 444–449.
114. Vogelstein, B., & Kinzler, K. W. (1999). Digital PCR. *Proceedings of the National Academy of Sciences of the United States of America, 96*, 9236–9241.
115. Bustin, S. A., & Nolan, T. (2004). Pitfalls of quantitative real-time reverse-transcription polymerase chain reaction. *Journal of Biomolecular Techniques: JBT, 15*, 155–166.
116. Li, X., Li, Y., Zhao, L., Zhang, D., Yao, X., Zhang, H., Wang, Y. C., Wang, X. Y., Xia, H., Yan, J., & Ying, H. (2014). Circulating muscle-specific miRNAs in Duchenne muscular dystrophy patients. *Molecular Therapy Nucleic acids, 3*, e177.
117. Heier, C. R., Fiorillo, A. A., Chaisson, E., Gordish-Dressman, H., Hathout, Y., Damsker, J. M., Hoffman, E. P., & Conklin, L. S. (2016). Identification of pathway-specific serum biomarkers of response to glucocorticoid and infliximab treatment in children with inflammatory Bowel disease. *Clinical and Translational Gastroenterology, 7*, e192.
118. Bak, R. O., Hollensen, A. K., Primo, M. N., Sorensen, C. D., & Mikkelsen, J. G. (2013). Potent microRNA suppression by RNA Pol II-transcribed 'Tough Decoy' inhibitors. *RNA, 19*, 280–293.
119. Hollensen, A. K., Bak, R. O., Haslund, D., & Mikkelsen, J. G. (2013). Suppression of microRNAs by dual-targeting and clustered tough decoy inhibitors. *RNA Biology, 10*, 406–414.
120. Ebert, M. S., Neilson, J. R., & Sharp, P. A. (2007). MicroRNA sponges: Competitive inhibitors of small RNAs in mammalian cells. *Nature Methods, 4*, 721–726.
121. Choi, W. Y., Giraldez, A. J., & Schier, A. F. (2007). Target protectors reveal dampening and balancing of Nodal agonist and antagonist by miR-430. *Science, 318*, 271–274.
122. Christopher, A. F., Kaur, R. P., Kaur, G., Kaur, A., Gupta, V., & Bansal, P. (2016). MicroRNA therapeutics: Discovering novel targets and developing specific therapy. *Perspectives in Clinical Research, 7*, 68–74.

Part III
Proteomic

Chapter 7
Proteomic Profiling of Human Skeletal Muscle in Health and Disease

Paul R. Langlais and Lawrence J. Mandarino

7.1 Theory and History of the Technique

Skeletal muscle is the largest organ in the body by mass [1] and as such can affect many processes in addition to its obvious roles in locomotion and breathing. In these roles, diseases of skeletal muscle can have profound effects on human health. Because it also is sensitive to the effect of insulin to enhance glucose uptake, storage, and metabolism, it also is a key organ in insulin resistance and deranged glucose metabolism in obesity and type 2 diabetes mellitus [2], one of the most common chronic diseases in the world. The complex polygenic nature of this disease, reaching the limits of predictive power of knowledge gained by genome-wide association studies [3], and the profound genotype/environment interaction required to develop type 2 diabetes mellitus compel additional approaches to understanding pathogenesis of the disease. These approaches by necessity, therefore, must be able to deal with complexities of biological changes in tissues. Proteomics offers investigators such an approach. Skeletal muscle is the largest single contributor when insulin stimulates clearance of glucose from the blood in humans [4] and is one of the major regulators of body amino acid metabolism [5]. Because of these and other vitally important functions, human skeletal muscle has been a widely studied organ, with nearly 20,000 citations in PubMed as of June 2018. Moreover, although not as accessible as blood, skeletal muscle can be sampled readily by the use of percutaneous needle biopsies taken under local anesthesia. Because of the low rate of adverse events during these procedures and the lack of lasting effects [6], muscle biopsies can readily and ethically be performed not only in patients in need of diagnosis of disease but also in healthy control volunteers. The highest incidence

P. R. Langlais · L. J. Mandarino (✉)
Division of Endocrinology, Department of Medicine, Center for Disparities in Diabetes, Obesity, and Metabolism, College of Medicine, University of Arizona Health Sciences, The University of Arizona, Tucson, AZ, USA
e-mail: mandarino@deptofmed.arizona.edu

J. G. Burniston, Y.-W. Chen (eds.), *Omics Approaches to Understanding Muscle Biology*, Methods in Physiology, https://doi.org/10.1007/978-1-4939-9802-9_7

of adverse events was for pain, approximately 1% of patients [6]. The use of percutaneous muscle biopsies in analyzing normal biology and disease pathogenesis dates to the nineteenth century but was popularized in modern medicine in the 1960s by Bergstrom [7], who described a muscle biopsy cannula that is still commonly used today. The history and methodological considerations of the use of the muscle biopsy technique have been reviewed thoroughly recently [8].

Skeletal muscle is a complex tissue in which multinucleated myotubes or myofibrils are organized into muscle fascicles that are surrounded by an extracellular matrix membrane (the perimysium) and contains blood vessels and sometimes infiltrating adipose tissue (Fig. 7.1). Although a needle or open muscle biopsy can contain these other tissues, when care is taken, a muscle biopsy predominantly represents muscle tissue. Therefore, results obtained from analysis of homogenates or lysates of a needle muscle biopsy can be assumed to primarily represent skeletal muscle. However, the possibility of confounding factors as a result of contamination of other tissues should always be kept in mind.

This chapter will focus on studies of human skeletal muscle using mass spectrometry-based proteomics technologies. The primary focus will be studies using biopsies to obtain skeletal muscle, but with selected references to studies using in vitro primary cultures of human myoblasts or differentiated myotubes. Although true in one sense or another for most tissues, because the primary function of skeletal muscle is to contract and do work, there are limitations to in vitro models. Moreover, with some exceptions, especially those of single gene mutations that produce disease, it is difficult to create animal models that adequately reflect humans. In complex, chronic diseases like type 2 diabetes, animal models in isolation offer limited insight. In addition, in studies that examine how exercise or exercise training affects protein abundance or other events in muscle, in vitro models mostly lack relevance because it is difficult to perform physiological muscle contraction in these models. Therefore, it is critical to continue to develop methods to study human skeletal muscle obtained directly using muscle biopsies. This chapter is not intended to be exhaustive, but rather to represent the array of areas which have proven to be amenable to mass spectrometry-based proteomics analysis in humans and for which such analyses have contributed in a major way to our base of knowledge.

Early studies of the human muscle proteome were limited by instrumentation. For example, one of the first studies of the human vastus lateralis proteome identified 107 proteins using a 2D gel/MS approach [9]. With the advent of Fourier transform ion cyclotron resonance mass spectrometry (FTICR/MS), the number of identified proteins grew to over 900 proteins [10] and Orbitrap-type instruments have expanded this to nearly 3000 proteins [11]. Moreover, earlier studies tended to be composed of lists of proteins, with little in the way of new biology. Later studies have focused on using mass spectrometry-based proteomics techniques to ask relevant biological questions and identify new proteins and posttranslational modifications. For example, studies have addressed how serine and threonine phosphorylation alters function of insulin signaling [12, 13], identification of binding partners of proteins [14], modification of ATP synthase beta subunit by phosphorylation

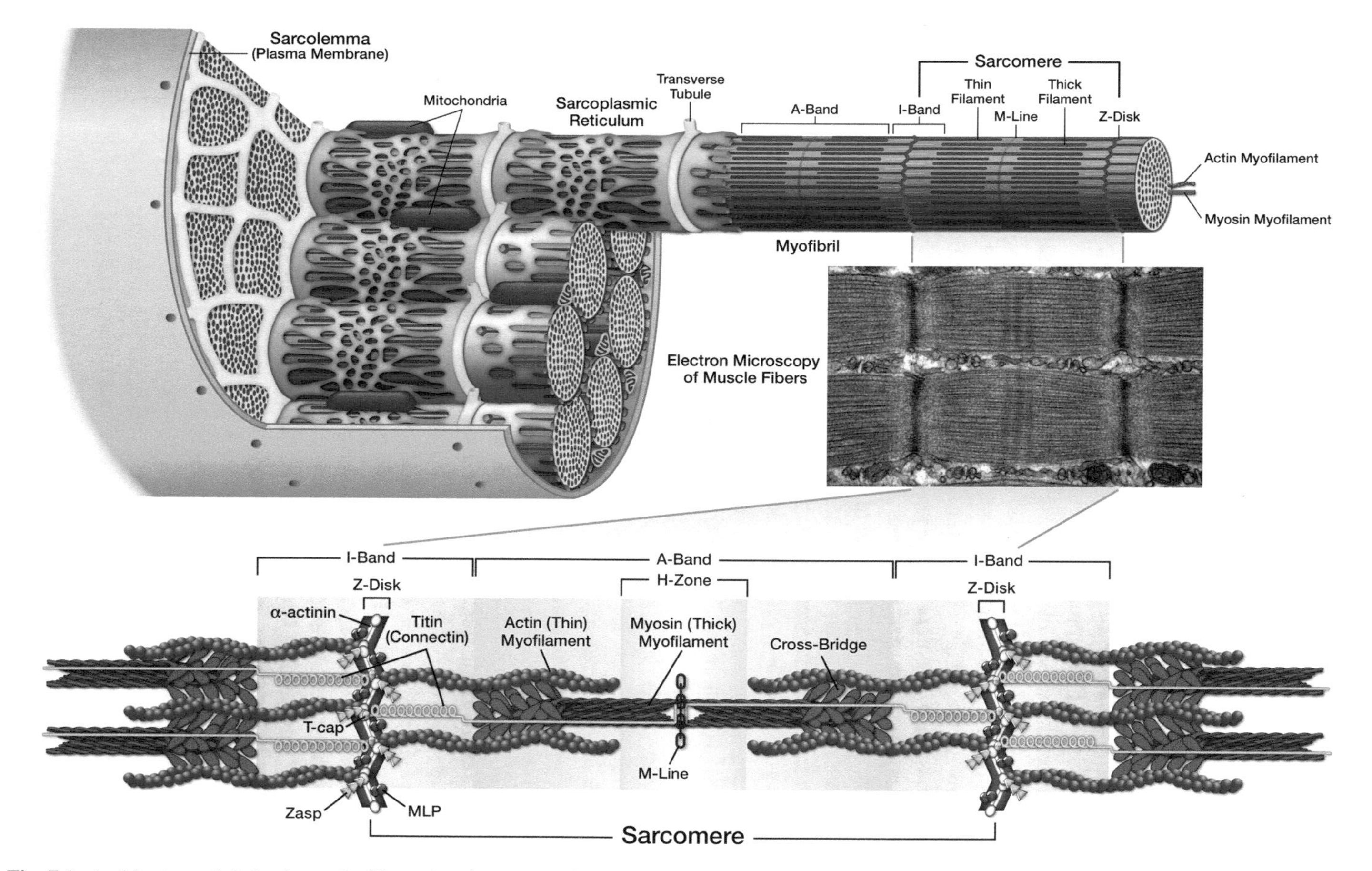

Fig. 7.1 Architecture of skeletal muscle, illustrating the structural organization of contractile proteins

[15, 16], how acetylation of mitochondrial proteins can affect their function [17], and others. In essence, mass spectrometry-based proteomics has grown from a more esoteric, instrument-focused approach to one in which highly sophisticated instruments are now able to be used by many investigators to ask questions in areas ranging from basic biology to translational research.

What advantages do study of the proteome confer? Analysis of genomic variation can provide information about the array of possibilities for biological variation in individuals and populations, but the genome must be transcribed to mRNA (and other forms of RNA) in order to express this variability. Since expression of mRNA is a highly regulated, complex, and gene-specific process, it is not possible at present to predict the extent of expression of any specific gene from the sequence of the genome. Despite this, there is a good overall relationship between mRNA expression and protein abundance [18]. Once mRNA is expressed, it must be translated to protein. Again, this is a highly regulated, complex process, the outcome of which is not possible to predict from genomic or mRNA analyses. Finally, once mRNA is translated into protein, there are many processes that control the abundance and activity of the protein. Included among these processes are posttranslational modifications of the protein. The advantages of mass spectrometry-based proteomics are that in a single experiment, quantitative information can be obtained about a large number of proteins on many levels. This includes information about protein isoforms, and in addition, posttranslational modifications can be identified and, in some cases, quantified. Thus, the use of proteomics complements and extends other forms of "-omics" data.

Modern mass spectrometry instrumentation has dramatically improved our ability to delve deeply into the proteome of all cells and tissues. In the past 20 years, the advances in instruments and techniques have seriously raised the bar for just how deep into the proteome investigators can dive. Earlier studies often used two-dimensional gel separation of proteins coupled with identification by mass spectrometry and quantification by protein staining intensity. These early studies often identified and quantified only a handful of proteins compared to more current methods but were not without merit. For example, because one of the dimensions of separation is isoelectric focusing, differentially phosphorylated (and charged) forms of the same proteins migrate with a characteristic pattern and allow investigators to obtain information they might not have gotten with other techniques, like immunoblots, at the time. An example of this would be the use of 2D gels to demonstrate that the beta subunit of mitochondrial ATP synthase exists in multiple phosphorylation states, a hitherto unappreciated example of phosphorylation of mitochondrial proteins [15]. More recently, investigators have used a one-dimensional "Gel-C-MS" approach (see below) or gel-free techniques with other fractionation techniques. Moreover, instrumentation has rapidly evolved over the past 20 years. It is not surprising, then, that coverage of the skeletal muscle proteome has increased dramatically over this period, from a handful of proteins to several hundred, to several thousand currently.

7.2 Major Applications

Because of its size and role in metabolism and locomotion, skeletal muscle often is a key tissue that is relevant to health and disease. This can result from a direct role of muscle in a genetic disease such as Duchenne muscular dystrophy, or indirectly, as in type 2 diabetes, through more subtle changes that lead to disease over time. And, of course, the primary function of skeletal muscle is to provide a means for the body to move through space, and the ability of muscle to respond to exercise therefore has been a key aspect of the study of skeletal muscle by mass spectrometry-based proteomics techniques. Proteomics techniques have been used in a vast array of studies of numerous processes in health and disease in human skeletal muscle. Examples of these studies to illustrate the breadth and depth of these investigations are given in Table 7.1.

Two of the most intensively studied areas that pertain to healthy skeletal muscle are exercise biology and aging. With regard to exercise and muscle contraction, the ability of skeletal muscle to adapt to stress is a key element of healthy skeletal muscle. The ability of muscle to respond to exercise may, in fact, be related to metabolic disease, with the evolution of the concept that insulin-resistant muscle also is "exercise resistant" [46, 47]. Proteomics techniques have played a key role in the study of adaptations to exercise in healthy muscle and in conditions such as obesity and type 2 diabetes mellitus. There has been an array of proteomics-based studies that have examined a range of questions related to exercise. Hostrup and colleagues used proteomics to study the question of whether treatment with beta-2 adrenergic agonists, such as those used for asthma treatment, can affect the adaptation of skeletal muscle to exercise training [19]. This question has relevance for athletes and others who use inhaled beta-2 agonists every day. Other investigators have used

Table 7.1 Areas of proteomic investigation in human skeletal muscle

Exercise	Hostrup et al. [19], Holloway et al. [20], Hussey et al. [21]
Aging	Baraibar et al. [22], Lourenco dos Santos et al. [23], Murgia et al. [24], Ohlendieck et al. [25], O'Connell et al. [26], Staunton et al. [27], Theron et al. [28]
Metabolism (obesity, type 2 diabetes, etc.)	Hwang et al. [29], Lefort et al. [30], Mielke et al. [17], Langlais et al. [12, 13], Hojlund et al. [10, 15, 16], Campbell et al. [11], Fiuza-Luces et al. [31]
Degenerative diseases	Conti et al. [32], Elf et al. [33], Greenberg et al. [34], Holland et al. [35], Mutsaers et al. [36], Salanova et al. [37], Sun et al. [38]
Ophthalmoplegia	Pfeffer et al. [39]
Inclusion body myositis	Guttsches et al. [40]
Lipid storage disorders	Debashree et al. [41]
Hypoxic stress	Capitanio et al. [42]
Growth hormone	Duran-Ortiz et al. [43]
Filaminopathy	Kley et al. [44]
Differentiation	Le Bihan et al. [45]

proteomics to define differences in the global proteome between exercise-trained and sedentary individuals and showed these people also differed in how their muscle responded to an acute bout of exercise [48]. Yet other investigators have used proteomics to study how exercise training influences protein abundance in a longitudinal manner. Burniston and colleagues used 2D gels, iTRAQ quantification, and LC-Maldi MS/MS to show that interval training resulted in broadly higher abundance of mitochondrial proteins. The 2D gel approach also allowed these investigators to examine differences in phosphorylation of creatine kinase and troponin T [48]. These results demonstrate the advantages of combining multiple proteomics techniques. Other investigators have used proteomics to study how exercise affects protein abundance in skeletal muscle of patients with type 2 diabetes, whose muscle is metabolically abnormal and characterized by reduced sensitivity to insulin [21]. These investigators could show that in patients with type 2 diabetes mellitus, only 4 weeks of exercise training could induce a rise in components of the malate–aspartate shuttle and citric acid cycle. Training also resulted in a fall in abundance of glycolytic proteins and changed the abundance of cytoskeleton proteins. These data showed that despite the overall lower response of insulin resistant muscle to a single exercise bout [46, 47], exercise training still conferred benefits in muscle of diabetic patients. With regard to studies of aging, again proteomics approaches have played an important part. These studies include those regarding changes in the neuromuscular junction in aging [49], carbonylation of proteins involved in glycolysis and protein quality control that may be involved in cellular senescence and other processes [22, 23], a shift to slow-twitch muscle fibers in aging and sarcopenia [24, 25], and aging in general [26–28].

A number of investigators have used proteomics methodology in an attempt to understand how differences in protein abundance between healthy and insulin resistant muscle in obese and type 2 diabetic patients might contribute to insulin resistance and the pathogenesis of type 2 diabetes mellitus. These studies have found similarities between insulin-resistant and sedentary muscle, especially in the mitochondrial proteome [29, 30]. This has led to the question of whether insulin resistance is a cause or consequence of lower mitochondrial content, capacity, or bioenergetic function. It is likely that this is a situation in which there is truth in both sides of this question. The observations that insulin-resistant muscle does not respond well to exercise [50–54] give some support to the idea that insulin resistance itself can lead to mitochondrial dysfunction, broadly defined. On the other hand, an improvement in insulin sensitivity with exercise training that raises mitochondrial abundance suggests that insulin resistance also can arise from low mitochondrial functional capacity. In yet another context, response to bariatric surgery, Campbell et al. provide a global analysis of protein abundance changes [11].

In addition to these well-studied areas, proteomics techniques have been used in a very broad range of studies of skeletal muscle biology in humans. Perhaps most widely represented are studies involving degenerative diseases of skeletal muscle (Table 7.1). These include genetic diseases such as Duchenne muscular dystrophy [35]; diseases that may have multiple causes such as amyotrophic lateral sclerosis, or Lou Gehrig's disease [32, 33]; facioscapulohumeral muscular dystrophy; and limb-girdle muscular dystrophy [34]. Other studies using proteomics have studied degenerative muscle diseases that result from denervation or disuse, such as bed rest [36–38]. Related to these diseases, one study used proteomics to study protein differences in ophthalmoplegia [39]. A number of other conditions that affect skeletal muscle also have been studied using proteomics techniques (Table 7.1). Included in this "miscellaneous" category are inclusion body myositis [40], lipid storage disorders [41], glycogen storage diseases [31], hypoxic stress [42], growth hormone excess and effects [43], filaminopathies [44], and normal differentiation of myocytes [45]. Here we provide more details on selected applications.

Quantification of Protein Abundance Using Biopsies of Human Skeletal Muscle In addition to assigning spectra to proteins and identifying a large number of proteins in the skeletal muscle proteome, it is desirable to quantify protein abundance, especially in the context of changes that occur in disease states or in healthy muscle in response to exercise and exercise training. When quantifying protein abundance from samples obtained in vivo, special considerations are required. In vitro, highly precise quantification of proteins can be accomplished efficiently with metabolic labeling of proteins, using techniques such as stable isotope labeling by amino acids in cell culture (SILAC) labeling with stable isotopes [55]. Although it is possible to isotopically label proteins in a mouse, it is not possible to do this in humans. This leaves two general options for those wishing to quantify protein abundance in muscle biopsies. One is chemical labeling of specific amino acid residues in proteins with a stable isotope, using such techniques as iTRAQ [20, 56]. This technique, while it can provide precise relative quantification of proteins, does not always yield the highest number of quantifiable proteins. It also is possible to quantify proteins without using labels, and this is termed label-free quantification. One method that has been used commonly is quantification by means of normalized spectral abundance factors [57–59]. In this technique, spectra specific to a given protein are summed, normalized to the length of the protein (longer proteins will have more spectra), and again normalized to the total spectrum count in a sample. The concept is that the number of spectra for a protein reflects its abundance, corrected for protein length and sample loading. This method has been used productively [29] but suffers from several limitations. First, if groups of separate samples are being compared, sometimes no spectra will be assigned to a low-abundance protein in some of the samples. This leads to "zero" abundance values and presents statistical normalization processes as well as the question of whether to use that sample at all. Second, the discontinuous nature of the data also presents analysis problems. To circumvent this, other investigators have used normalized peptide ion intensities to quantify proteins. This technique also has issues if a particular protein is not assigned to be statistically significantly present in a particular sample. Some analysis programs attempt to circumvent this problem by comparing two-dimensional "peptide ion feature" maps that can aid in assigning low-abundance peaks to a particular peptide and protein. While no method is perfect, and it is difficult to provide absolute, stoichiometric protein quantification with in vivo samples, important conclusions can still be derived from many of these relative quantification methods.

Proteomes of Subcellular Fractions of Skeletal Muscle: Mitochondria Despite the difficulty of performing clean isolations of subcellular fractions of skeletal muscle biopsies, a number of studies have attempted to provide these proteomes. Because production of energy for cellular functions such as contraction is a key aspect of the biology of skeletal muscle in health and disease, the mitochondrial fraction has attracted attention. One of the earlier attempts to define the human skeletal muscle mitochondrial proteome used percutaneous needle muscle biopsies from healthy volunteers and a 1D gel-based approach to identify over 800 unique proteins, of which nearly 500 were annotated to the mitochondrion at that time [60]. This proportion demonstrates the extent of contamination of mitochondrial subfractions in skeletal muscle, since the investigators isolated these mitochondria using a technique designed to produce functionally active organelles [30, 60]. Although, on a protein basis, more than a third of the proteins assigned appeared to be contaminants of the mitochondrial compartment, the proportion of spectra assigned to mitochondrial proteins was much higher. Regardless of these limitations, this approach allows investigators to study the proteome of an organelle in the same preparation that can be used for functional studies.

Other Subcellular Proteomes Although much attention has been paid to the mitochondrial fraction of skeletal muscle, in particular due to interest in the biology of exercise as well as skeletal muscle metabolism (both discussed below), other subcellular proteomes have received attention. Some of these proteomes have been obtained using muscle biopsies, while others derive from studies using cultured cells. For example, one study used a variety of techniques, including proteomics, to study the molecular nature of the neuromuscular junction in rodents and humans [49]. This study revealed similarities and differences between rodent and human neuromuscular junctions. Other studies, probably necessarily, used cultured myocytes to obtain other information on muscle-related proteomes. Studies of the secretome are especially amenable to the in vitro approach and would be difficult if not impossible to perform in vivo. In one study, primary cultures of human skeletal muscle cells were used for this purpose [61]. The purpose of this study was to identify potential myokines, proteins secreted by skeletal muscle cells that can act on other tissues at a distance. Using this approach, the investigators identified 12 novel potential myokines [61]. This study is a prime example of how results from unbiased proteomics analyses can lay the ground for generating hypotheses and potentially discovering new biology.

Analysis of Posttranslational Modifications Protein function can be modulated drastically by posttranslational modifications. Early discoveries by Sutherland of the regulation of glycogen phosphorylase and glycogen metabolism by phosphorylation in response to hormonal stimulation are the classical example of how protein phosphorylation regulates activity and was described in a series of classical papers [62–65]. Posttranslational modifications have been a long-standing specialty of mass spectrometry, beginning with identifying which specific residues within a protein are modified and evolving to quantification of directional changes of thousands of posttranslational modifications across entire proteomes. When working on whole proteomes and phosphorylation, after enzymatic digestion of the protein sample, the resulting peptide mixture is overwhelmingly non-phosphorylated, so much so that phosphopeptides become undetectable due to relatively low abundance. However, this limitation is commonly overcome by employing phosphopeptide enrichment, a technique that has evolved to become standard practice. Tandem mass spectrometry, which takes advantage of coupling a parent ion scan to a peptide ion fragmentation scan, has proven to be an excellent tool for assigning which exact amino acid residue is subject to posttranslational modification within a given peptide. Soon after tandem mass spectrometry became a mainstream go-to for peptide and posttranslational modification site identification, the ability to perform quantification of directional changes in peptide and posttranslational modification abundance between different experimental groups emerged, eventually resulting in a powerful technique and a new field, quantitative proteomics. Experiments evaluating directional fold changes in posttranslational modifications like acetylation and phosphorylation across entire proteomes are performed with the aforementioned label and label-free techniques, and new oversight practices that allow for better review of data quality are now requiring investigators to submit all annotated posttranslationally modified peptide spectra to data repositories. With respect to proteomic analysis of skeletal muscle and regulation of protein function by phosphorylation, there are a number of examples. One of the more complicated examples of such analyses was the mapping and quantification of serine and threonine phosphorylation sites in the insulin signaling protein insulin receptor substrate (IRS)-1. Tyrosine phosphorylation of IRS-1 by the tyrosine kinase activity of the insulin receptor transduces the signal from insulin to increase glucose uptake and metabolism in tissues like skeletal muscle and adipose tissue, and a great deal of attention was paid to the potential role that might be played by this protein in the pathophysiology of skeletal muscle insulin resistance in type 2 diabetes and obesity (see below). Extensive studies of IRS-1 in cell culture and rodent models from the laboratory of Morris White revealed that in addition to positive modulation of IRS-1 function by tyrosine phosphorylation, serine and threonine phosphorylation of IRS-1 also could regulate its function [66, 67]. Because of the importance of insulin resistance to the development of and consequences from type 2 diabetes, and since human diabetes differs substantially in a variety of respects from cell culture or animal models of the disease, a compelling case could be made for determining how IRS-1 is phosphorylated in human muscle obtained from biopsies. Several phospho-specific antibodies were developed to quantify the extent of phosphorylation at a few biologically relevant serine and threonine sites in IRS-1, for use in immunoblots. Although these antibodies work well in cell culture, the immunoblots from lysates of human muscle biopsies were often somewhat unclear. Moreover, because approximately 270 of the 1242 amino acid residues of IRS-1 are serine and threonine residues, it would be very difficult to use immunoblot analysis to obtain a global picture of IRS-1 phosphorylation in human muscle. As techniques for analysis and quantification of serine/threonine sites in proteins by mass spectrometry were developed, investigators applied them to analysis of IRS-1 in human muscle biopsies in order to circumvent these problems. In a series of papers, a number of serine/threonine phosphorylation sites were identified and quantified under basal and insulin-stimulated conditions, where possible, in IRS-1 present at endogenous levels in human muscle biopsies [12, 13, 68–70], and found a number of differences in phosphorylation of serine and threonine residues in insulin-resistant muscle. These studies also revealed new biology related to human IRS-1, which differs slightly but significantly from mouse IRS-1 [47], in addition to providing information relevant to disease processes.

Another posttranslational modification of proteins that is amenable to analysis by mass spectrometry methods is lysine acetylation [71–75]. Lysine residues often confer function to proteins through their positive charge, and acetylation of lysine removes this positive charge. Therefore, lysine acetylation can dramatically alter the function of a protein. Acetylation of lysine residues is overrepresented in metabolic and mitochondrial proteins [73], perhaps because of an environment in which acetyl groups are especially abundant. An unbiased acetylomic analysis of human skeletal muscle proteins revealed a number of acetylation sites in mitochondrial proteins [17]. Among these were three lysine acetylation sites in the ADP/ATP translocase 1, sometimes referred to as adenine nucleotide translocase 1 (ANT1). In examining the crystal structure of bovine ANT1, it became apparent that one of these residues, lysine 23, was part of a cluster of positively charged residues near the adenine nucleotide-binding pocket. Molecular modeling of the acetylated and unacetylated versions of ANT1 revealed that acetylation at lysine 23 could dramatically lower the binding affinity of ANT1 for ADP [17]. Since ADP is a key respiratory signal and because ANT1 is an important locus of control of respiration under conditions of low to moderate energy demand, a modification at this single residue might dramatically affect cell function [76]. Moreover, the extent of acetylation at lysine 23 was related to insulin sensitivity in skeletal muscle, demonstrating a biomedically relevant role for this single acetylation site. It is likely that more lysine acetylation sites on proteins in human skeletal muscle, and other tissues, will be identified by mass spectrometry analysis and will lead to studies of the functional relevance of acetylation. Parenthetically, although acetyltransferases and deacetylases have been identified in particular instances, it is still unclear whether enzymatic or nonenzymatic acetylation is involved in mitochondrial proteins, and nonenzymatic acetylation may be widespread [77].

7.3 Limitations

Skeletal muscle is a highly organized, complex tissue whose main function is to provide movement and do work. The arrangement of contractile proteins and mitochondria within muscle is shown in Fig. 7.1. Contraction is accomplished via the interaction of myosin and skeletal muscle actin, with titin providing springlike elasticity. Proteins of the Z-disk, such as alpha actinin-2, provide an anchor for alpha-actin and titin. This highly structured molecular arrangement yields the typical appearance of the sarcomere, the functional unit of contraction (Fig. 7.1). Mitochondria are abundant and embedded between myofibrils, making it harder to liberate functional mitochondria than it is in less organized cell types, such as hepatocytes. Because of this organizational complexity and specialized function, skeletal muscle, especially biopsies taken from humans or muscle specimens taken from other animals, presents some special difficulties for many proteomics analyses. First, because of the abundance of connective tissue and the complexity of its structure consisting of "sticky" contractile proteins, this tissue requires vigorous disruption to solubilize a large number of proteins. This also makes it more difficult to isolate relatively pure subcellular fractions if one wishes to focus on a particular proteome, for example, the mitochondrial or nuclear proteome. With regard to mitochondria, if the experiment calls for isolation of functionally active mitochondria in order to study bioenergetics of skeletal muscle, one must be willing to accept less-than-pure mitochondria preparations from the perspective of proteomics [30, 60]. The implication of that is that the sensitivity and comprehensive nature of mass spectrometry-based proteomics techniques allow one to "see" all the contamination that is present normally in preparations of subcellular organelles like mitochondria. But perhaps the biggest methodological hurdle to overcome in proteomics analysis of skeletal muscle is the very high abundance of a few large to very large proteins. Myosin, titin, and skeletal muscle actin comprise a very large proportion of the total protein in the myocyte, as can be appreciated from Fig. 7.1. Therefore, techniques using in-solution enzymatic digests of muscle biopsy lysates, as opposed to gel slices, are plagued by a poor ability to identify low-abundance proteins. This is due in part to the fact that current instruments have a limit on the number of spectra that can be analyzed in any given time interval. Although scanning speeds have improved dramatically in the last 10 years, as cited above, the problem in analyzing skeletal muscle lysates is still severe. This problem can be likened to trying to find blades of grass that are hidden in a vast forest of trees. One continues to identify trees. Similarly, with these techniques, one continually identifies large contractile and structural proteins (the trees) and misses lower-abundance proteins like transcription factors, enzymes, and signaling proteins (the leaves of grass). Therefore, it is advantageous to employ techniques that use molecular weight as one dimension of a multidimensional fractionation protocol designed to maximize the ability to assign proteins in the muscle proteome. Perhaps the most efficient way to do this is to perform what has been termed "Gel-C-MS." A typical workflow for this approach is shown in Fig. 7.2. A complex lysate of a skeletal muscle biopsy first is resolved using a one-dimensional SDS–polyacrylamide gel. This concentrates large proteins at the top of the gel. Very large proteins like titin will not even enter the gel and thus will be primarily excluded. Although myosin will enter most gels commonly used for these purposes, the myosin band is easily identified by its characteristic staining with Coomassie blue and can be totally avoided if desired. Each lane of the gel is then cut into 6–12 slices, and each slice is enzymatically digested and processed for analysis [10]. This procedure, with current instruments, can yield over 2000 quantifiable proteins from a single, individual skeletal muscle biopsy and in upward of

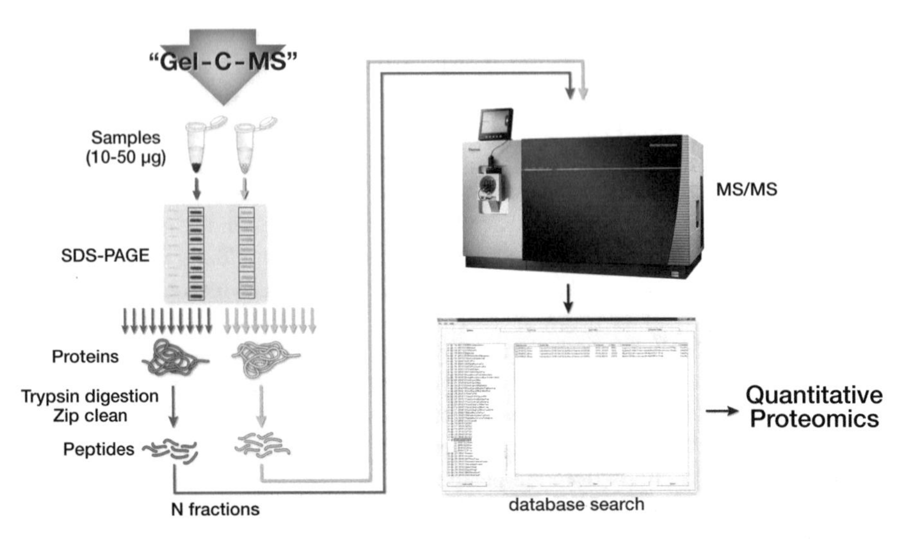

Fig. 7.2 Typical workflow for "Gel-C-MS" for proteomics analysis of lysates of skeletal muscle biopsies

3000 quantifiable proteins over a number of specimens in any given experiment. This discrepancy exists because not all the same proteins are assigned in all subjects in a study. The problem of interference from large contractile proteins is not as severe with myoblasts or even differentiated myotubes studied in culture, but if human skeletal muscle is to be studied in a physiological or pathophysiological context, these problems must be faced.

Another consideration when designing experiments may be whether one wishes to study the effects of acute or chronic muscle contraction or exercise. For these studies, the in vivo approach in humans, using muscle biopsies, may be the best. Although there are now approaches to electrically stimulate differentiated myotubes in culture, it is not clear how well this approach compares with muscle contraction in vivo. For example, in vivo exercise produces effects not only through contraction itself but also in response to systemic physiological changes that occur during exercise, such as changes in blood flow and hormone or substrate concentrations. The in vivo approach integrates these changes. It also is possible, using one-legged muscle contraction [78], to minimize some of these systemic changes and relatively isolate the effects of "muscle contraction" from those of "exercise."

7.4 Vital Future Directions

From early lists of proteins generated on now-outmoded instrumentation to the use of modern instruments that have the capability to dive deep into the proteome, the field already has achieved significant progress. At some point soon, instrumentation should allow us to identify and quantify (at least relatively) almost all proteins that are expressed in human muscle. Where do we go from here? It is critical that mass spectrometry-based proteomics move to its place as a technique accessible to many investigators who want to use its power to ask directed questions about basic or translational biology. Currently, it too often is still a technique shrouded in mystery behind a proteomics "core," where investigators have little knowledge or control over the data that are generated. Proteomics facilities must move from service cores to collaborative laboratories engaged in generating data in a transparent, interactive manner. It is also vital that methods be developed to more accurately and precisely quantify proteins, using either a label-free or labeling approach.

Acknowledgments The authors greatly acknowledge the artistic expertise of Marv Ruona. This work was supported in part by R01DK047936 (LJM), R01DK066483 (LJM), and R01DK098493 (PRL).

References

1. Pedersen, B. K., & Febbraio, M. A. (2012). Muscles, exercise and obesity: Skeletal muscle as a secretory organ. *Nature Reviews. Endocrinology, 8*, 457–465.
2. Defronzo, R. A. (2009). Banting lecture. From the triumvirate to the ominous octet: A new paradigm for the treatment of type 2 diabetes mellitus. *Diabetes, 58*, 773–795.
3. Vassy, J. L., Hivert, M. F., Porneala, B., Dauriz, M., Florez, J. C., Dupuis, J., Siscovick, D. S., Fornage, M., Rasmussen-Torvik, L. J., Bouchard, C., & Meigs, J. B. (2014). Polygenic type 2 diabetes prediction at the limit of common variant detection. *Diabetes, 63*, 2172–2182.
4. DeFronzo, R. A., Jacot, E., Jequier, E., Maeder, E., Wahren, J., & Felber, J. P. (1981). The effect of insulin on the disposal of intravenous glucose. Results from indirect calorimetry and hepatic and femoral venous catheterization. *Diabetes, 30*, 1000–1007.
5. Wolfe, R. R. (2006). The underappreciated role of muscle in health and disease. *The American Journal of Clinical Nutrition, 84*, 475–482.
6. Neves, M., Jr., Barreto, G., Boobis, L., Harris, R., Roschel, H., Tricoli, V., Ugrinowitsch, C., Negrao, C., & Gualano, B. (2012). Incidence of adverse events associated with percutaneous muscular biopsy among healthy and diseased subjects. *Scandinavian Journal of Medicine & Science in Sports, 22*, 175–178.
7. Bergstrom, J. (1962). Muscle electrolytes in man. *Scandinavian Journal of Clinical and Laboratory Investigation, 14*, 100–110.
8. Bergstrom, J. (1962). Laryngologic aspects of the treatment of acute barbiturate poisoning. *Acta Oto-Laryngologica. Supplementum, 173*, 1–59.
9. Gelfi, C., De Palma, S., Cerretelli, P., Begum, S., & Wait, R. (2003). Two-dimensional protein map of human vastus lateralis muscle. *Electrophoresis, 24*, 286–295.
10. Hojlund, K., Yi, Z., Hwang, H., Bowen, B., Lefort, N., Flynn, C. R., Langlais, P., Weintraub, S. T., & Mandarino, L. J. (2008). Characterization of the human skeletal muscle proteome by one-dimensional gel electrophoresis and HPLC-ESI-MS/MS. *Molecular & Cellular Proteomics, 7*, 257–267.
11. Campbell, L. E., Langlais, P. R., Day, S. E., Coletta, R. L., Benjamin, T. R., De Filippis, E. A., Madura, J. A., 2nd, Mandarino, L. J., Roust, L. R., & Coletta, D. K. (2016). Identification of novel changes in human skeletal muscle proteome after Roux-en-Y gastric bypass surgery. *Diabetes, 65*, 2724–2731.

12. Langlais, P., Mandarino, L. J., & Yi, Z. (2010). Label-free relative quantification of co-eluting isobaric phosphopeptides of insulin receptor substrate-1 by HPLC-ESI-MS/MS. *Journal of the American Society for Mass Spectrometry, 21*, 1490–1499.
13. Langlais, P., Yi, Z., Finlayson, J., Luo, M., Mapes, R., De Filippis, E., Meyer, C., Plummer, E., Tongchinsub, P., Mattern, M., & Mandarino, L. J. (2011). Global IRS-1 phosphorylation analysis in insulin resistance. *Diabetologia, 54*, 2878–2889.
14. Geetha, T., Langlais, P., Luo, M., Mapes, R., Lefort, N., Chen, S. C., Mandarino, L. J., & Yi, Z. (2011). Label-free proteomic identification of endogenous, insulin-stimulated interaction partners of insulin receptor substrate-1. *Journal of the American Society for Mass Spectrometry, 22*, 457–466.
15. Hojlund, K., Wrzesinski, K., Larsen, P. M., Fey, S. J., Roepstorff, P., Handberg, A., Dela, F., Vinten, J., McCormack, J. G., Reynet, C., & Beck-Nielsen, H. (2003). Proteome analysis reveals phosphorylation of ATP synthase beta -subunit in human skeletal muscle and proteins with potential roles in type 2 diabetes. *The Journal of Biological Chemistry, 278*, 10436–10442.
16. Hojlund, K., Yi, Z., Lefort, N., Langlais, P., Bowen, B., Levin, K., Beck-Nielsen, H., & Mandarino, L. J. (2010). Human ATP synthase beta is phosphorylated at multiple sites and shows abnormal phosphorylation at specific sites in insulin-resistant muscle. *Diabetologia, 53*, 541–551.
17. Mielke, C., Lefort, N., McLean, C. G., Cordova, J. M., Langlais, P. R., Bordner, A. J., Te, J. A., Ozkan, S. B., Willis, W. T., & Mandarino, L. J. (2014). Adenine nucleotide translocase is acetylated in vivo in human muscle: Modeling predicts a decreased ADP affinity and altered control of oxidative phosphorylation. *Biochemistry, 53*, 3817–3829.
18. Yi, Z., Bowen, B. P., Hwang, H., Jenkinson, C. P., Coletta, D. K., Lefort, N., Bajaj, M., Kashyap, S., Berria, R., De Filippis, E. A., & Mandarino, L. J. (2008). Global relationship between the proteome and transcriptome of human skeletal muscle. *Journal of Proteome Research, 7*, 3230–3241.
19. Hostrup, M., Onslev, J., Jacobson, G. A., Wilson, R., & Bangsbo, J. (2018). Chronic β2 -adrenoceptor agonist treatment alters muscle proteome and functional adaptations induced by high intensity training in young men. *The Journal of Physiology, 596*, 231–252.
20. Holloway, K. V., O'Gorman, M., Woods, P., Morton, J. P., Evans, L., Cable, N. T., Goldspink, D. F., & Burniston, J. G. (2009). Proteomic investigation of changes in human vastus lateralis muscle in response to interval-exercise training. *Proteomics, 9*, 5155–5174.
21. Hussey, S. E., Sharoff, C. G., Garnham, A., Yi, Z., Bowen, B. P., Mandarino, L. J., & Hargreaves, M. (2013). Effect of exercise on the skeletal muscle proteome in patients with type 2 diabetes. *Medicine and Science in Sports and Exercise, 45*, 1069–1076.
22. Baraibar, M., Hyzewicz, J., Rogowska-Wrzesinska, A., Bulteau, A. L., Prip-Buus, C., Butler-Browne, G., & Friguet, B. (2014). Impaired metabolism of senescent muscle satellite cells is associated with oxidative modifications of glycolytic enzymes. *Free Radical Biology & Medicine, 75*(Suppl 1), S23.
23. Lourenco dos Santos, S., Baraibar, M. A., Lundberg, S., Eeg-Olofsson, O., Larsson, L., & Friguet, B. (2015). Oxidative proteome alterations during skeletal muscle ageing. *Redox Biology, 5*, 267–274.
24. Murgia, M., Toniolo, L., Nagaraj, N., Ciciliot, S., Vindigni, V., Schiaffino, S., Reggiani, C., & Mann, M. (2017). Single muscle Fiber proteomics reveals Fiber-type-specific features of human muscle aging. *Cell Reports, 19*, 2396–2409.
25. Ohlendieck, K. (2011). Proteomic profiling of fast-to-slow muscle transitions during aging. *Frontiers in Physiology, 2*, 105.
26. O'Connell, K., Gannon, J., Doran, P., & Ohlendieck, K. (2007). Proteomic profiling reveals a severely perturbed protein expression pattern in aged skeletal muscle. *International Journal of Molecular Medicine, 20*, 145–153.
27. Staunton, L., Zweyer, M., Swandulla, D., & Ohlendieck, K. (2012). Mass spectrometry-based proteomic analysis of middle-aged vs. aged vastus lateralis reveals increased levels of carbonic

anhydrase isoform 3 in senescent human skeletal muscle. *International Journal of Molecular Medicine, 30*, 723–733.
28. Theron, L., Gueugneau, M., Coudy, C., Viala, D., Bijlsma, A., Butler-Browne, G., Maier, A., Bechet, D., & Chambon, C. (2014). Label-free quantitative protein profiling of vastus lateralis muscle during human aging. *Molecular & Cellular Proteomics, 13*, 283–294.
29. Hwang, H., Bowen, B. P., Lefort, N., Flynn, C. R., De Filippis, E. A., Roberts, C., Smoke, C. C., Meyer, C., Hojlund, K., Yi, Z., & Mandarino, L. J. (2010). Proteomics analysis of human skeletal muscle reveals novel abnormalities in obesity and type 2 diabetes. *Diabetes, 59*, 33–42.
30. Lefort, N., Glancy, B., Bowen, B., Willis, W. T., Bailowitz, Z., De Filippis, E. A., Brophy, C., Meyer, C., Hojlund, K., Yi, Z., & Mandarino, L. J. (2010). Increased reactive oxygen species production and lower abundance of complex I subunits and carnitine palmitoyltransferase 1B protein despite normal mitochondrial respiration in insulin-resistant human skeletal muscle. *Diabetes, 59*, 2444–2452.
31. Fiuza-Luces, C., Santos-Lozano, A., Llavero, F., Campo, R., Nogales-Gadea, G., Diez-Bermejo, J., Baladron, C., Gonzalez-Murillo, A., Arenas, J., Martin, M. A., Andreu, A. L., Pinos, T., Galvez, B. G., Lopez, J. A., Vazquez, J., Zugaza, J. L., & Lucia, A. (2018). Muscle molecular adaptations to endurance exercise training are conditioned by glycogen availability: A proteomics-based analysis in the McArdle mouse model. *The Journal of Physiology, 596*, 1035–1061.
32. Conti, A., Riva, N., Pesca, M., Iannaccone, S., Cannistraci, C. V., Corbo, M., Previtali, S. C., Quattrini, A., & Alessio, M. (2014). Increased expression of myosin binding protein H in the skeletal muscle of amyotrophic lateral sclerosis patients. *Biochimica et Biophysica Acta, 1842*, 99–106.
33. Elf, K., Shevchenko, G., Nygren, I., Larsson, L., Bergquist, J., Askmark, H., & Artemenko, K. (2014). Alterations in muscle proteome of patients diagnosed with amyotrophic lateral sclerosis. *Journal of Proteomics, 108*, 55–64.
34. Greenberg, S. A., Salajegheh, M., Judge, D. P., Feldman, M. W., Kuncl, R. W., Waldon, Z., Steen, H., & Wagner, K. R. (2012). Etiology of limb girdle muscular dystrophy 1D/1E determined by laser capture microdissection proteomics. *Annals of Neurology, 71*, 141–145.
35. Holland, A., Carberry, S., & Ohlendieck, K. (2013). Proteomics of the dystrophin-glycoprotein complex and dystrophinopathy. *Current Protein & Peptide Science, 14*, 680–697.
36. Mutsaers, C. A., Lamont, D. J., Hunter, G., Wishart, T. M., & Gillingwater, T. H. (2013). Label-free proteomics identifies Calreticulin and GRP75/Mortalin as peripherally accessible protein biomarkers for spinal muscular atrophy. *Genome Medicine, 5*, 95.
37. Salanova, M., Gelfi, C., Moriggi, M., Vasso, M., Vigano, A., Minafra, L., Bonifacio, G., Schiffl, G., Gutsmann, M., Felsenberg, D., Cerretelli, P., & Blottner, D. (2014). Disuse deterioration of human skeletal muscle challenged by resistive exercise superimposed with vibration: Evidence from structural and proteomic analysis. *The FASEB Journal, 28*, 4748–4763.
38. Sun, H., Qiu, J., Chen, Y., Yu, M., Ding, F., & Gu, X. (2014). Proteomic and bioinformatic analysis of differentially expressed proteins in denervated skeletal muscle. *International Journal of Molecular Medicine, 33*, 1586–1596.
39. Pfeffer, G., Gorman, G. S., Griffin, H., Kurzawa-Akanbi, M., Blakely, E. L., Wilson, I., Sitarz, K., Moore, D., Murphy, J. L., Alston, C. L., Pyle, A., Coxhead, J., Payne, B., Gorrie, G. H., Longman, C., Hadjivassiliou, M., McConville, J., Dick, D., Imam, I., Hilton, D., Norwood, F., Baker, M. R., Jaiser, S. R., Yu-Wai-Man, P., Farrell, M., McCarthy, A., Lynch, T., McFarland, R., Schaefer, A. M., Turnbull, D. M., Horvath, R., Taylor, R. W., & Chinnery, P. F. (2014). Mutations in the SPG7 gene cause chronic progressive external ophthalmoplegia through disordered mitochondrial DNA maintenance. *Brain, 137*, 1323–1336.
40. Guttsches, A. K., Brady, S., Krause, K., Maerkens, A., Uszkoreit, J., Eisenacher, M., Schreiner, A., Galozzi, S., Mertens-Rill, J., Tegenthoff, M., Holton, J. L., Harms, M. B., Lloyd, T. E., Vorgerd, M., Weihl, C. C., Marcus, K., & Kley, R. A. (2017). Proteomics of rimmed vacuoles define new risk allele in inclusion body myositis. *Annals of Neurology, 81*, 227–239.

41. Debashree, B., Kumar, M., Keshava Prasad, T. S., Natarajan, A., Christopher, R., Nalini, A., Bindu, P. S., Gayathri, N., & Srinivas Bharath, M. M. (2018). Mitochondrial dysfunction in human skeletal muscle biopsies of lipid storage disorder. *Journal of Neurochemistry, 145*, 323–341.
42. Capitanio, D., Fania, C., Torretta, E., Vigano, A., Moriggi, M., Bravata, V., Caretti, A., Levett, D. Z. H., Grocott, M. P. W., Samaja, M., Cerretelli, P., & Gelfi, C. (2017). TCA cycle rewiring fosters metabolic adaptation to oxygen restriction in skeletal muscle from rodents and humans. *Scientific Reports, 7*, 9723.
43. Duran-Ortiz, S., Brittain, A. L., & Kopchick, J. J. (2017). The impact of growth hormone on proteomic profiles: A review of mouse and adult human studies. *Clinical Proteomics, 14*, 24.
44. Kley, R. A., Maerkens, A., Leber, Y., Theis, V., Schreiner, A., van der Ven, P. F., Uszkoreit, J., Stephan, C., Eulitz, S., Euler, N., Kirschner, J., Muller, K., Meyer, H. E., Tegenthoff, M., Furst, D. O., Vorgerd, M., Muller, T., & Marcus, K. (2013). A combined laser microdissection and mass spectrometry approach reveals new disease relevant proteins accumulating in aggregates of filaminopathy patients. *Molecular & Cellular Proteomics, 12*, 215–227.
45. Le Bihan, M. C., Barrio-Hernandez, I., Mortensen, T. P., Henningsen, J., Jensen, S. S., Bigot, A., Blagoev, B., Butler-Browne, G., & Kratchmarova, I. (2015). Cellular proteome dynamics during differentiation of human primary myoblasts. *Journal of Proteome Research, 14*, 3348–3361.
46. De Filippis, E., Alvarez, G., Berria, R., Cusi, K., Everman, S., Meyer, C., & Mandarino, L. J. (2008). Insulin-resistant muscle is exercise resistant: Evidence for reduced response of nuclear-encoded mitochondrial genes to exercise. *American Journal of Physiology. Endocrinology and Metabolism, 294*, E607–E614.
47. McLean, C. S., Mielke, C., Cordova, J. M., Langlais, P. R., Bowen, B., Miranda, D., Coletta, D. K., & Mandarino, L. J. (2015). Gene and MicroRNA expression responses to exercise; relationship with insulin sensitivity. *PLoS One, 10*, e0127089.
48. Schild, M., Ruhs, A., Beiter, T., Zugel, M., Hudemann, J., Reimer, A., Krumholz-Wagner, I., Wagner, C., Keller, J., Eder, K., Kruger, K., Kruger, M., Braun, T., Niess, A., Steinacker, J., & Mooren, F. C. (2015). Basal and exercise induced label-free quantitative protein profiling of m. vastus lateralis in trained and untrained individuals. *Journal of Proteomics, 122*, 119–132.
49. Jones, R. A., Harrison, C., Eaton, S. L., Llavero Hurtado, M., Graham, L. C., Alkhammash, L., Oladiran, O. A., Gale, A., Lamont, D. J., Simpson, H., Simmen, M. W., Soeller, C., Wishart, T. M., & Gillingwater, T. H. (2017). Cellular and molecular anatomy of the human neuromuscular junction. *Cell Reports, 21*, 2348–2356.
50. Dela, F., & Stallknecht, B. (2010). Effect of physical training on insulin secretion and action in skeletal muscle and adipose tissue of first-degree relatives of type 2 diabetic patients. *American Journal of Physiology. Endocrinology and Metabolism, 299*, E80–E91.
51. Hernandez-Alvarez, M. I., Thabit, H., Burns, N., Shah, S., Brema, I., Hatunic, M., Finucane, F., Liesa, M., Chiellini, C., Naon, D., Zorzano, A., & Nolan, J. J. (2010). Subjects with early-onset type 2 diabetes show defective activation of the skeletal muscle PGC-1{alpha}/Mitofusin-2 regulatory pathway in response to physical activity. *Diabetes Care, 33*, 645–651.
52. Nilsson, M. I., Greene, N. P., Dobson, J. P., Wiggs, M. P., Gasier, H. G., Macias, B. R., Shimkus, K. L., & Fluckey, J. D. (2010). Insulin resistance syndrome blunts the mitochondrial anabolic response following resistance exercise. *American Journal of Physiology. Endocrinology and Metabolism, 299*, E466–E474.
53. Roberts, M. D., Kerksick, C. M., Dalbo, V. J., Hassell, S. E., Tucker, P. S., & Brown, R. (2010). Molecular attributes of human skeletal muscle at rest and after unaccustomed exercise: An age comparison. *Journal of Strength and Conditioning Research, 24*, 1161–1168.
54. Stephens, N. A., & Sparks, L. M. (2015). Resistance to the beneficial effects of exercise in type 2 diabetes: Are some individuals programmed to fail? *The Journal of Clinical Endocrinology and Metabolism, 100*, 43–52.

55. Ong, S. E., Blagoev, B., Kratchmarova, I., Kristensen, D. B., Steen, H., Pandey, A., & Mann, M. (2002). Stable isotope labeling by amino acids in cell culture, SILAC, as a simple and accurate approach to expression proteomics. *Molecular & Cellular Proteomics, 1*, 376–386.
56. Wiese, S., Reidegeld, K. A., Meyer, H. E., & Warscheid, B. (2007). Protein labeling by iTRAQ: A new tool for quantitative mass spectrometry in proteome research. *Proteomics, 7*, 340–350.
57. Pavelka, N., Fournier, M. L., Swanson, S. K., Pelizzola, M., Ricciardi-Castagnoli, P., Florens, L., & Washburn, M. P. (2008). Statistical similarities between transcriptomics and quantitative shotgun proteomics data. *Molecular & Cellular Proteomics, 7*, 631–644.
58. Zybailov, B., Mosley, A. L., Sardiu, M. E., Coleman, M. K., Florens, L., & Washburn, M. P. (2006). Statistical analysis of membrane proteome expression changes in *Saccharomyces cerevisiae*. *Journal of Proteome Research, 5*, 2339–2347.
59. Zybailov, B. L., Florens, L., & Washburn, M. P. (2007). Quantitative shotgun proteomics using a protease with broad specificity and normalized spectral abundance factors. *Molecular BioSystems, 3*, 354–360.
60. Lefort, N., Yi, Z., Bowen, B., Glancy, B., De Filippis, E. A., Mapes, R., Hwang, H., Flynn, C. R., Willis, W. T., Civitarese, A., Hojlund, K., & Mandarino, L. J. (2009). Proteome profile of functional mitochondria from human skeletal muscle using one-dimensional gel electrophoresis and HPLC-ESI-MS/MS. *Journal of Proteomics, 72*, 1046–1060.
61. Hartwig, S., Raschke, S., Knebel, B., Scheler, M., Irmler, M., Passlack, W., Muller, S., Hanisch, F. G., Franz, T., Li, X., Dicken, H. D., Eckardt, K., Beckers, J., de Angelis, M. H., Weigert, C., Haring, H. U., Al-Hasani, H., Ouwens, D. M., Eckel, J., Kotzka, J., & Lehr, S. (2014). Secretome profiling of primary human skeletal muscle cells. *Biochimica et Biophysica Acta, 1844*, 1011–1017.
62. Berthet, J., Rall, T. W., & Sutherland, E. W. (1957). The relationship of epinephrine and glucagon to liver phosphorylase. IV. Effect of epinephrine and glucagon on the reactivation of phosphorylase in liver homogenates. *Journal of Biological Chemistry, 224*, 463–475.
63. Rall, T. W., Sutherland, E. W., & Wosilait, W. D. (1956). The relationship of epinephrine and glucagon to liver phosphorylase. III. Reactivation of liver phosphorylase in slices and in extracts. *Journal of Biological Chemistry, 218*, 483–495.
64. Sutherland, E. W., & Wosilait, W. D. (1956). The relationship of epinephrine and glucagon to liver phosphorylase. I. Liver phosphorylase; preparation and properties. *The Journal of Biological Chemistry, 218*, 459–468.
65. Wosilait, W. D., & Sutherland, E. W. (1956). The relationship of epinephrine and glucagon to liver phosphorylase. II. Enzymatic inactivation of liver phosphorylase. *Journal of Biological Chemistry, 218*, 469–481.
66. Aguirre, V., Uchida, T., Yenush, L., Davis, R., & White, M. F. (2000). The c-Jun NH(2)-terminal kinase promotes insulin resistance during association with insulin receptor substrate-1 and phosphorylation of Ser(307). *The Journal of Biological Chemistry, 275*, 9047–9054.
67. Aguirre, V., Werner, E. D., Giraud, J., Lee, Y. H., Shoelson, S. E., & White, M. F. (2002). Phosphorylation of Ser307 in insulin receptor substrate-1 blocks interactions with the insulin receptor and inhibits insulin action. *Journal of Biolgical Chemistry, 277*, 1531–1537.
68. Luo, M., Langlais, P., Yi, Z., Lefort, N., De Filippis, E. A., Hwang, H., Christ-Roberts, C. Y., & Mandarino, L. J. (2007). Phosphorylation of human insulin receptor substrate-1 at serine 629 plays a positive role in insulin signaling. *Endocrinology, 148*, 4895–4905.
69. Luo, M., Reyna, S., Wang, L., Yi, Z., Carroll, C., Dong, L. Q., Langlais, P., Weintraub, S. T., & Mandarino, L. J. (2005). Identification of insulin receptor substrate 1 serine/threonine phosphorylation sites using mass spectrometry analysis: Regulatory role of serine 1223. *Endocrinology, 146*, 4410–4416.
70. Yi, Z., Langlais, P., De Filippis, E. A., Luo, M., Flynn, C. R., Schroeder, S., Weintraub, S. T., Mapes, R., & Mandarino, L. J. (2007). Global assessment of regulation of phosphorylation of insulin receptor substrate-1 by insulin in vivo in human muscle. *Diabetes, 56*, 1508–1516.

71. Choudhary, C., Kumar, C., Gnad, F., Nielsen, M. L., Rehman, M., Walther, T. C., Olsen, J. V., & Mann, M. (2009). Lysine acetylation targets protein complexes and co-regulates major cellular functions. *Science, 325*, 834–840.
72. Hirschey, M. D., Shimazu, T., Goetzman, E., Jing, E., Schwer, B., Lombard, D. B., Grueter, C. A., Harris, C., Biddinger, S., Ilkayeva, O. R., Stevens, R. D., Li, Y., Saha, A. K., Ruderman, N. B., Bain, J. R., Newgard, C. B., Farese, R. V., Jr., Alt, F. W., Kahn, C. R., & Verdin, E. (2010). SIRT3 regulates mitochondrial fatty-acid oxidation by reversible enzyme deacetylation. *Nature, 464*, 121–125.
73. Wang, Q., Zhang, Y., Yang, C., Xiong, H., Lin, Y., Yao, J., Li, H., Xie, L., Zhao, W., Yao, Y., Ning, Z. B., Zeng, R., Xiong, Y., Guan, K. L., Zhao, S., & Zhao, G. P. (2010). Acetylation of metabolic enzymes coordinates carbon source utilization and metabolic flux. *Science, 327*, 1004–1007.
74. Zhang, J., Sprung, R., Pei, J., Tan, X., Kim, S., Zhu, H., Liu, C. F., Grishin, N. V., & Zhao, Y. (2009). Lysine acetylation is a highly abundant and evolutionarily conserved modification in *Escherichia coli. Molecular & Cellular Proteomics, 8*, 215–225.
75. Zhao, S., Xu, W., Jiang, W., Yu, W., Lin, Y., Zhang, T., Yao, J., Zhou, L., Zeng, Y., Li, H., Li, Y., Shi, J., An, W., Hancock, S. M., He, F., Qin, L., Chin, J., Yang, P., Chen, X., Lei, Q., Xiong, Y., & Guan, K. L. (2010). Regulation of cellular metabolism by protein lysine acetylation. *Science, 327*, 1000–1004.
76. Willis, W. T., Miranda-Grandjean, D., Hudgens, J., Willis, E. A., Finlayson, J., De Filippis, E. A., Zapata Bustos, R., Langlais, P. R., Mielke, C., & Mandarino, L. J. (2018). Dominant and sensitive control of oxidative flux by the ATP-ADP carrier in human skeletal muscle mitochondria: Effect of lysine acetylation. *Archives of Biochemistry and Biophysics, 647*, 93–103.
77. Kulkarni, R. A., Worth, A. J., Zengeya, T. T., Shrimp, J. H., Garlick, J. M., Roberts, A. M., Montgomery, D. C., Sourbier, C., Gibbs, B. K., Mesaros, C., Tsai, Y. C., Das, S., Chan, K. C., Zhou, M., Andresson, T., Weissman, A. M., Linehan, W. M., Blair, I. A., Snyder, N. W., & Meier, J. L. (2017). Discovering targets of non-enzymatic acylation by thioester reactivity profiling. *Cell Chemical Biology, 24*, 231–242.
78. Helge, J. W., Stallknecht, B., Richter, E. A., Galbo, H., & Kiens, B. (2007). Muscle metabolism during graded quadriceps exercise in man. *The Journal of Physiology, 581*, 1247–1258.

Chapter 8
Proteome Profiling of Muscle Cells and Muscle Tissue Using Stable Isotope Labeling by Amino Acids

Emily Canessa, Mansi V. Goswami, Alison M. Samsel, Michael Ogundele, Shefa M. Tawalbeh, Tchilabalo D. Alayi, and Yetrib Hathout

8.1 Theory and History of Stable Isotope Labeling by Amino Acids for Proteomic Studies

Comprehensive and accurate proteome profiling of skeletal muscle remains challenging owing to the large proteome dynamic range in this tissue and the increased sensitivity needed to detect low-abundant proteins. Sarcomeric and glycolytic enzymes are by far the most abundant proteins in muscle, masking detection and quantification of low-abundant proteins such as dystrophin, dystrophin-associated protein complex, cell signaling proteins, and transcription factors. About 5400 unique proteins have been identified so far in skeletal muscle using extensive pre-fractionation methods and mass spectrometry [1]. While this is good for cataloging the muscle proteome, pre-fractionation methods are often not compatible with quantification or comparative proteomics because of inherent technical variability from experiment to experiment leading to false and inaccurate quantification. To overcome these challenges, several stable isotope labeling strategies have been developed and tested in the past in different cell culture models and tissues. Typically, the two samples to be compared are tagged with heavy and light stable isotope-labeled moieties, respectively, either at the peptide level, after digestion of proteins with a protease (e.g., trypsin or endoproteinase Lys-C), or even at the cellular level before protein extraction and processing. These different stable isotope labeling techniques are described elsewhere [2], and for this book chapter, we will focus on the most accurate methods.

E. Canessa · M. V. Goswami · A. M. Samsel · M. Ogundele · S. M. Tawalbeh · T. D. Alayi · Y. Hathout (✉)
School of Pharmacy and Pharmaceutical Sciences, Binghamton University – SUNY, Johnson City, NY, USA
e-mail: ecanessa@binghamton.edu; mgoswami@binghamton.edu; asamsel@binghamton.edu; mogunde1@binghamton.edu; stawalb1@binghamton.edu; talayi@binghamton.edu; yhathout@binghamton.edu

J. G. Burniston, Y.-W. Chen (eds.), *Omics Approaches to Understanding Muscle Biology*, Methods in Physiology, https://doi.org/10.1007/978-1-4939-9802-9_8

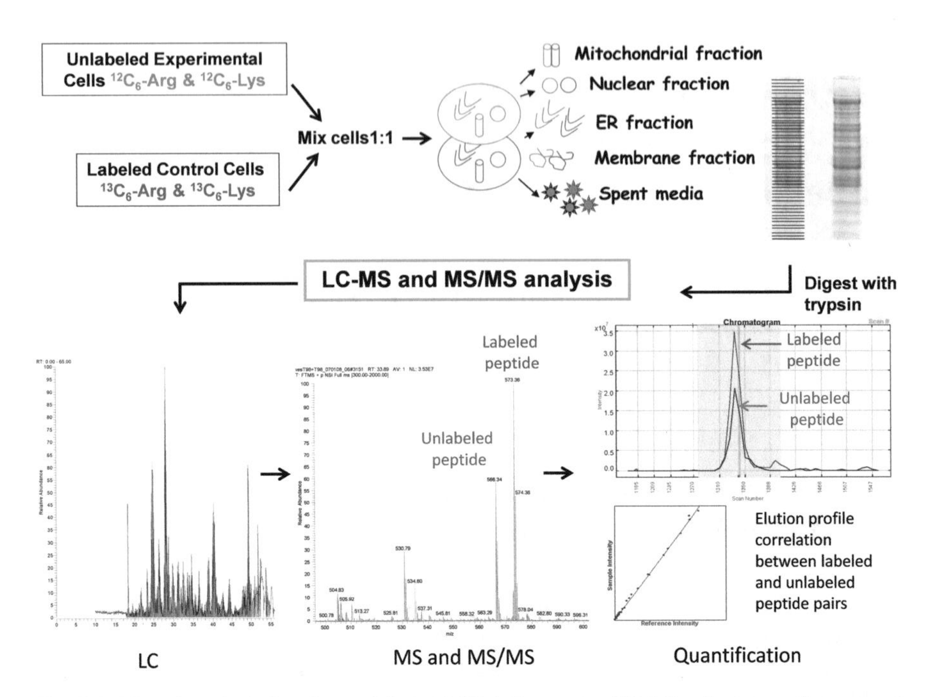

Fig. 8.1 Flowchart depicting the workflow of SILAC strategy. SILAC cells (most often control cells labeled with $^{13}C_6$-Lys and $^{13}C_6$-Arg) and unlabeled cells (most often experimental cells grown in $^{12}C_6$,$^{14}N_2$-Lys and $^{12}C_6$-Arg) are mixed at a 1:1 ratio for subsequent subcellular fractionation. Total proteins are extracted from each fraction of interest and further fractionated by SDS-PAGE to reduce complexity. Gel lanes are then sliced into 2–3 mm pieces. Each piece is in-gel digested by trypsin and the resulting peptides analyzed by LC-MS/MS. Labeled (red) and unlabeled (blue) peptide pairs co-elute, but their masses are resolved by mass spectrometry. The intensity ratio of unlabeled to labeled peptide reflects the amount of the corresponding protein in experimental relative to control cells. The bottom right panel shows the elution profile of labeled and unlabeled peptide pair

8.1.1 Stable Isotope Labeling by Amino Acid in Cell Culture (SILAC)

Stable isotope labeling by amino acid in cell culture (SILAC) is by far the most accurate method for quantitative proteomics when dealing with a cell culture system. SILAC was first introduced in 2002 using light unlabeled leucine (Leu-d0) and heavy deuterium labeled leucine (Leu-d3) to measure temporal changes in the expression of proteins over the course of C2C12 differentiation [3]. The SILAC technique was further improved by incorporating dual labeling using $^{13}C_6$-Arg and $^{13}C_6$-Lys, the two amino acids at which trypsin cuts. As depicted in Fig. 8.1, one set of cell culture, often control cells, is grown in media where normal Arg and Lys are replaced by heavy $^{13}C_6$-Arg and $^{13}C_6$-Lys, respectively, while the experimental cells are grown in classical unlabeled media with normal $^{12}C_6$-Arg and $^{12}C_6$-Lys. After passaging the cells five times (equivalent to seven cell doublings) in their respective media, the proteome of SILAC control cells becomes fully labeled with heavy Arg and Lys residues, while the proteome of the experimental cells contains normal light Arg and Lys. A given tryptic peptide belonging to a given protein will have the exact same sequence in stable isotope-labeled control cells as its unlabeled homologue in the experimental cells. Typically, total protein extracts from SILAC labeled cells and unlabeled cells are mixed at a 1:1 ratio. The mixture is then digested by trypsin and the resulting peptides are analyzed on a reversed-phase liquid chromatography connected to a tandem mass spectrometry system with MS and MS/MS capabilities (LC-MS/MS). During chromatography, stable isotope labeled and unlabeled peptide pairs found in the same protein will co-elute since they have exactly the same amino acid sequence, and therefore the same hydrophobicity but their masses are resolved by the MS detector since the stable isotope-labeled peptide is 6 Da heavier than the unlabeled peptide. The MS/MS data of either labeled or unlabeled peptide is used to identify the protein, while the elution profile of labeled and unlabeled peptide pairs detected at the MS level is used to determine the ratio of that protein in experimental samples relative to the SILAC labeled control.

Because the light and heavy stable isotope-labeled sample pairs are mixed before protein digestion and sometimes even before subcellular fractionation and protein extraction (Fig. 8.1), variation due to sample handling and processing is minimized. Typical precision or CV using this technology is below 15%. The SILAC strategy has been widely used in a number of eloquent studies ranging from basic comparative proteomics [4] to more complex proteomics studies such as quantitative subcellular proteome profiling [5, 6], protein-protein interactions [7, 8], phosphoproteomics and cell signaling [9, 10]. As discussed below SILAC has been also implemented in studies using muscle cell culture as a model.

8.1.2 Stable Isotope Labeling in Mammals (SILAM)

More recently SILAC strategy has been extended to animal models where the whole animal proteome is labeled with a stable isotope-labeled essential amino acid such as $^{13}C_6$-Lys [11] or uniformly labeled with ^{15}N [12]. In this strategy, pregnant female mice are given $^{13}C_6$-Lys labeled mouse feed or ^{15}N-labeled Spirulina-based feed and the entire mouse colony is maintained under this regimen for a couple of generations. All proteins in the tissue and body fluids of the second-generation mice become fully labeled with the heavy Lys residue or ^{15}N. The proteome of the SILAM mouse is labeled with $^{13}C_6$-Lys and mixed at 1:1 ratio to the proteomes of unlabeled control and experimental mice. In this example only tryptic peptides that contain Lys residues can be used for quantification. But when the proteome of the SILAM mouse is labeled with ^{15}N and spiked into the proteomes of unlabeled control and experimental mice, all tryptic peptides can be used for quantification since all peptides contain several nitrogen atoms. In the SILAM $^{13}C_6$-Lys experiment, the mass difference between stable isotope-labeled peptide and its unlabeled peptide homologue is 6 Da, while in the SILAM ^{15}N experiment, the mass difference

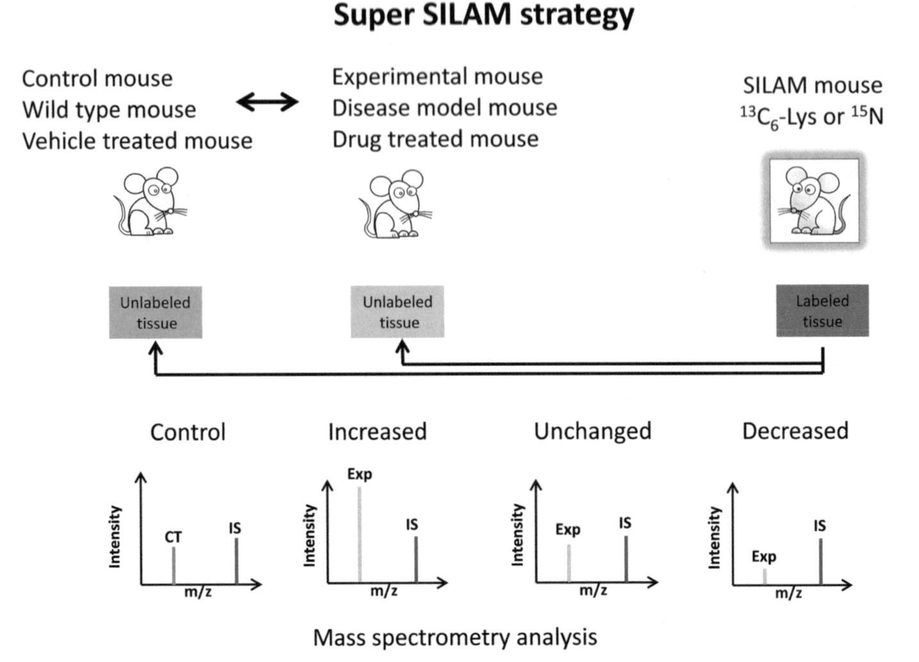

Fig. 8.2 Flowchart depicting the workflow of super SILAM strategy. Protein extracts from tissue or body fluids of experimental and control mice are spiked with known amount of protein extract from matching tissue or body fluids. Spiked samples are fractionated by SDS-PAGE and processed for in-gel digestion and mass spectrometry analysis. Depending on the requested experiments, replicate mice are used ($n = 3$ to 6 mice per group). Proteins are then identified and quantified using our IP2 software. The lower panel shows an example of a peptide from a protein that was 1:1 ratio in control (blue) relative to internal standard (red). Then a scenario where a peptide from a protein was 2:1 ratio, 1:1 ratio, or 1:2 ratio in experimental mice (green) relative to internal standard (red). Note that the amount of spiked internal standard is the same for this peptide of interest, allowing for accurate comparisons of protein levels across different samples

between stable isotope-labeled peptide and its unlabeled homologue depends on the number of nitrogen atoms in the peptide. In general, when using SILAM strategy, protein extract from a given tissue or body fluid from the SILAM mouse is used as an internal standard to quantify the same set of proteins in unlabeled experimental and control mice using intensity ratios of heavy and light peptide pairs (Fig. 8.2).

These innovative in vivo labeling strategies utilizing $^{13}C_6$-Lys and ^{15}N were successfully implemented in a number of impactful research studies. This includes a study on the role of specific genes using knockout mouse models [13], one defining alteration in cell signaling during skin carcinogenesis [14], one defining disease altered protein turnover in the brain of Alexander disease mouse model [15], studies on the development of the brain in rat models [16, 17], as well as a study on proteome turnover in different tissues and organs [18].

8.2 Implementation of SILAC and SILAM Strategies to Study the Proteome of Muscle in Health and Disease Conditions

8.2.1 In Vitro Studies Using SILAC Labeled Muscle Cell Cultures

Because of its accuracy and specificity, SILAC strategies have been implemented in a number of basic and translational studies including studies dealing with cultured muscle cell models. We highlight below some of the major applications in the field.

8.2.1.1 Study of Muscle Cell Differentiation and Myogenesis

One of the first applications of SILAC strategy was actually designed to define changes in protein expression during muscle cell differentiation using C2C12 cells as a model [3]. In this initial study, the author cultured undifferentiated C2C12 myoblasts in media containing unlabeled Leu-d0, while C2C12 undergoing differentiation were grown in media containing Leu-d3 and were harvested at days 0, 2, and 5 over the course of differentiation. Total protein extracts from Leu-d3 labeled C2C12 day 0, day 2, and day 5 were mixed at 1:1 ratio with total protein extract from undifferentiated Leu-d0 C2C12 and analyzed by LC-MS/MS. The authors identified several proteins that did not change over the course of differentiation, while other proteins either increased or decreased. This early study was performed using C2C12 cells, a transformed and immortalized mouse muscle cell line, and was done for only 5 days of differentiation.

To gain a better idea of how myogenesis occurs in human muscle, a recent study used human primary myoblast in combination with a triple SILAC strategy to define changes in the proteome dynamic during muscle cell differentiation [19]. Because the authors used light, medium, and heavy stable isotope-labeled Lys and Arg, as well as a more recent instrument, they identified key proteins involved in myogenesis but also unraveled several novel proteins never reported before. The authors identified 243 proteins that significantly changed during differentiation and grouped them into 5 different clusters: those that decreased upon initiating differentiation (clusters 1 and 3) and those that increased during differentiation (clusters 4 and 5). Cluster 2 rose at day 1 of differentiation and then returned to normal levels. This is one of the most comprehensive studies of the timecourse of protein changes during myogenesis in vitro and an excellent example of how versatile the SILAC strategy can be.

8.2.1.2 Study of the Muscle Cell Secretome

It is well accepted that muscle tissue secretes a large number of biologically active proteins such as myokines which regulate the secretion of a number of other biologically active molecules that have a key role in metabolism [20]. It is challenging to study the muscle secretome in vivo owing to difficulty in accessing the interstitial space between muscle fibers. Thus, growing differentiated myotube cell cultures was a first step in cataloguing the muscle cell secretome. In the past few years, a number of muscle cell secretome studies have published a list of proteins that were detected in the conditioned media of cultured myoblast and myotubes using C2C12 cells or primary human muscle cells [21–23]. Besides myokines, differentiated muscle cells secrete a large number of other biologically relevant proteins. About 554–954 nonredundant proteins have been identified so far in the conditioned media of these cultured myotubes. Only 25% of these proteins had a signal peptide and were secreted through the classical pathway [23]. The remaining 75% of proteins in the conditioned media were either released by leaking cells or via exosomes [23]. Some of the key secreted proteins identified in the conditioned media of differentiated myotubes are listed in Table 8.1 with their Swiss-Prot accession number and potential function relevant to muscle.

A key challenge in studying cell secretome is distinguishing between proteins that are truly released by the cells and protein contaminants that originate from the 2% bovine serum added in differentiation culture media or gelatins used to coat culture flasks. Indeed, several bovine sera proteins have high sequence homology with human or mouse proteins and might confound the true origin of the proteins identified in the conditioned media. However, SILAC strategy can overcome this background noise and distinguish proteins that are truly synthesized and secreted by SILAC labeled cells from proteins that might be contaminants which are not labeled by heavy isotope $^{13}C_6$-Arg or $^{13}C_6$-Lys residues.

Our group implemented SILAC strategy to define differential exosomal proteins secreted between dystrophin deficient mouse muscle cells and healthy normal mouse muscle cells [24]. This study demonstrated that dystrophin-deficient muscle cells exhibit disrupted vesicle trafficking and protein secretion.

8.2.2 *In Vivo Studies Using SILAM Labeled Mouse Model for Muscle Diseases*

8.2.2.1 Use of SILAM Strategy to Define Alterations in the Proteome of Dystrophin-Deficient Skeletal Muscle

SILAM was successfully implemented by our group to elucidate molecular events involved in muscle pathogenesis in the dystrophin-deficient mouse model mdx-52 [25] and in a mouse model of inflammatory myopathy [25]. Using this strategy, thousands of proteins were identified and 789 were quantified, allowing for their

Table 8.1 Partial list of proteins identified in the conditioned media of differentiated human primary myotubes (selected from [23])

Accession number	Protein name	Potential function relevant to skeletal muscle
Q15063	Periostin	Extracellular matrix (ECM) organization and response to muscle activity
P15502	Elastin	Skeletal muscle development
Q14118	Dystroglycan	Sarcolemma anchoring and skeletal muscle tissue regeneration
P35052	Glypican-1	Positive regulation of skeletal muscle cell differentiation
P07585	Decorin	Skeletal muscle tissue development
O14793	Myostatin or GDF8	Negative regulation of muscle hypertrophy, negative regulation of myoblast proliferation and differentiation
Q12841	Follistatin-related protein 1	A myokine involved in muscle fiber regeneration
P01137	TGFB1	Negative regulation of skeletal muscle tissue development
Q15389	Angiopoietin-1	Regulation of skeletal muscle satellite cell proliferation
P17936	IGFBP3	Positive regulation of myoblast differentiation
Q99988	GDF15	Positive regulation of myoblast fusion
P78504	Protein jagged-1	Myoblast differentiation
P25391	Laminin	Involved in myoblast fusion
P09382	Galectin-1	Involved in myoblast fusion
P09486	SPARC	Positively regulated in ECM remodeling
P08253	Matrix metalloproteinase-2	Positively regulated in ECM remodeling
P02751	Fibronectin	Collagen-/actin-binding protein, ECM remodeling
P08123	Collagen alpha-2 (I) chain	Upregulated as differentiation occurs, ECM remodeling
P55291	Cadherin-15	Positive regulation of muscle cell differentiation
P19022	Cadherin-2	Positive regulation of muscle cell differentiation

levels of expression to be compared between muscle disease mouse models and wild-type mice. Figure 8.3 shows the dystrophin protein which is absent in the skeletal muscle of an mdx mouse model compared to that of wild-type mouse, GAPDH (glyceraldehyde-3-phosphate dehydrogenase) which remains unchanged, and talin which is significantly elevated in the skeletal muscle of a dystrophin-lacking mouse. Detailed data about altered levels in the proteome of skeletal dystrophin-deficient skeletal muscle obtained using SILAM is discussed elsewhere [25, 26]. The key finding from this SILAM mouse study was the identification of pathways involved in early-stage muscular dystrophy pathogenesis in the mdx mouse model. A subset of these findings were confirmed in human muscle tissue comparing DMD muscle tissue to healthy muscle tissue [25].

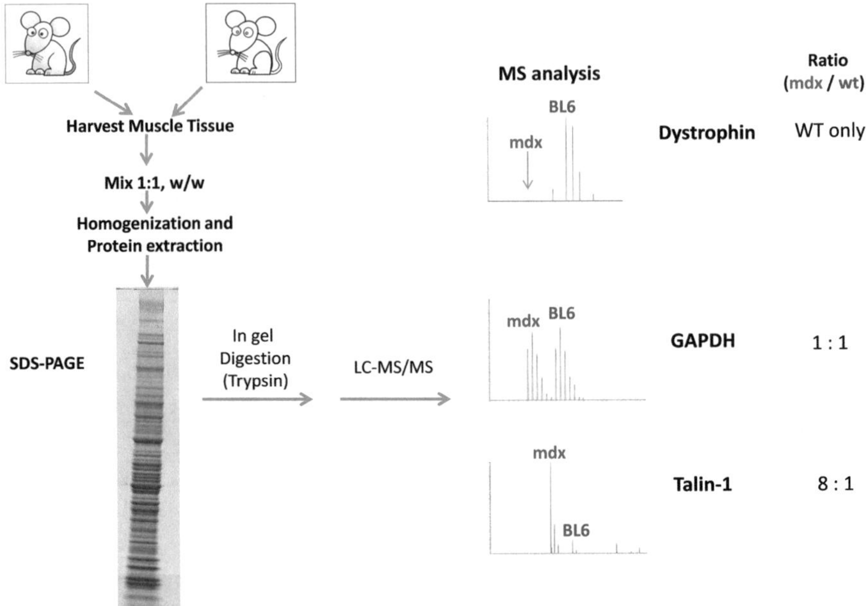

Fig. 8.3 Principle of SILAM mouse strategy to define the proteome signature of dystrophin-deficient skeletal muscle in the mdx mouse model. Protein extract from the muscle tissue of a SILAM labeled wild-type mouse is mixed at 1:1 ratio with protein extract from the muscle tissue of an mdx mouse. The protein mixture is then fractionated by gel electrophoresis. Bands are excised and in-gel digested by trypsin. The resulting peptides are analyzed by LC-MS/MS. Thousands of proteins were identified and quantified. Right panel shows examples of light and heavy peptide pairs detected for dystrophin, GAPDH, and talin-1 in mdx and SILAM wild type, respectively. Using the intensity ratio of light to heavy peptide will determine the ratio of corresponding protein in the muscle of mdx mouse relative to wild-type mouse

8.2.2.2 Use of SILAM Strategy to Accurately Quantify Levels of Dystrophin Protein in Skeletal Muscle

The need for accurate and precise dystrophin quantification in muscle biopsies and skeletal tissue collected from clinical and preclinical studies has become highly important since the introduction of dystrophin replacement therapies. Examples of such therapies are exon skipping using a phosphorodiamidate morpholino oligomer (PMO) [27], codon read through using ataluren drug, gene therapy using micro-dystrophin AAV vector, and stem cell therapy [28]. While these therapies hold promise because they repair the primary defect in DMD (e.g., restoration of the missing dystrophin in the skeletal muscle), the precise amount and function of the restored dystrophin remain a challenge to define which often delays approval by regulatory agencies and halts further clinical trials. Indeed, quantification of

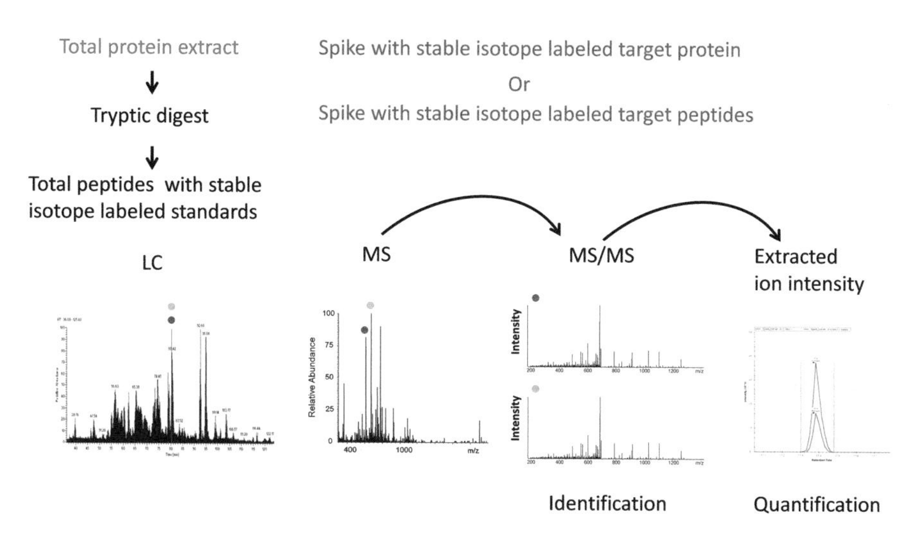

Fig. 8.4 Targeted mass spectrometry assay principle and flowchart. The quantification is done with MS/MS transition ions rather than *m/z* values at the MS level. This increases the sensitivity and specificity of the measurement because MS/MS data from modern high-resolution instruments often have little or no background noise. Quantification at the MS levels on anther hand can be noisier with a risk of false quantification due to co-eluting peptides that might have a very similar mass to the target peptide. Targeted mass spectrometry uses a combination of peptide (MS) selection followed by quantification of a "daughter" fragment ion (MS/MS). This greatly improves the specificity of the measurement because the likely hood of a co-eluting peptide having an identical fragment ion is low

dystrophin using classical antibody-based assays such as Western blot and immunofluorescence exhibited large variability when done on the same biopsies across different laboratories and even within the same laboratory [29]. Furthermore, different results can be obtained when using immunostaining depending on the muscle section, the antibodies, and image reading [30].

Stable isotope-labeled proteotypic peptide(s) can be used as internal standard(s) to quantify any given protein in a complex biological sample using targeted mass spectrometry methods. Figure 8.4 depicts the targeted mass spectrometry flowchart. Basically, the method focuses on selecting a peptide or peptides of interest based on their retention time and *m/z* value in a LC-MS/MS run. Then the sample to be analyzed is spiked with a known amount of stable isotope peptides that have the same sequences as the unlabeled peptides in the target protein. This allows quantification of the target peptides and therefore allows us to quantify the amount of protein of interest in the sample. The stable isotope peptides could be either custom synthesized and spiked into the sample after total protein digestion or generated from full-length stable isotope-labeled protein that is spiked into the sample before digestion with trypsin. The full-length stable isotope protein could be generated by SILAC cultured cells that express the protein of interest or by using tissue samples from SILAM mouse that express the protein of interest.

We have successfully used this method to accurately quantify dystrophin in human muscle biopsies using full-length $^{13}C_6$-Lys labeled mouse dystrophin as an internal standard [31]. Indeed, since mouse dystrophin and human dystrophin share 91% sequence homology, several tryptic peptides had exactly the same sequence between human and mouse. These peptides were selected and used for quantification of dystrophin.

8.2.2.3 Use of SILAM Mouse to Define Serum Protein Biomarkers Associated with Dystrophin Deficiency

Fully ^{15}N-labeled wild-type mice were successfully used as surrogates to identify and quantify serum protein biomarkers associated with dystrophin deficiency using two independent mouse models, the engineered mdx-Δ52, and the naturally occurring mdx-23 [32]. Of the 355 screened mouse sera proteins, 23 were found significantly elevated and 4 significantly lowered in sera samples of mdx mice relative to wild-type mice (p value <0.001). Elevated proteins were mostly of muscle origin and included myofibrillar proteins, glycolytic enzymes, transport proteins, and other muscle proteins, while decreased proteins were mostly of extracellular origin. Furthermore, analysis of sera from 1 week to 7 months old mdx-23 mice revealed age-dependent changes in the level of these biomarkers. Most biomarkers acutely elevated at 3 weeks of age and then either remained increased or decreased in older mdx mice up to 7 months old. This data was validated in sera samples of DMD patients [32]. This study shows the utility of using SILAM mouse to define serum circulating biomarkers that are associated with muscle disease and could be extended to a number of other applications when looking at the response of these biomarkers.

8.3 Limitations and Extensions of the Technique

While the use of SILAC and SILAM strategies coupled with mass spectrometry enables accurate proteome profiling of cultured cells and animal tissues, respectively, these techniques have some limitations as described below.

SILAC and SILAM strategies suffer from low throughput because only pairs of samples can be compared at a time. To overcome some of the throughput challenges, investigators have introduced super SILAC strategy, where a SILAC labeled cell batch is prepared in large quantities and used as internal standard to quantify a series of experimental cells [33]. Similarly, super SILAM can be used in the same way to perform proteome profiling of different experimental mouse models. With improved sample preparation methods and state-of-the-art LC-MS/MS instruments with higher speed, resolution, and sensitivity, the quantitative coverage of the proteome will dramatically improve and several samples can be processed in only 1 week instead of a month. But sample preparation still remains the limiting step in a number of proteomic studies, especially those implementing SILAC and SILAM strategies, because it takes on average 1 month to fully label cells with SILAC and up to 6 months to generate a SILAM mouse colony.

Another limitation of SILAC/SILAM strategies is they are restricted for use only in cell culture models and animal models. Although this is true, cultured SILAC labeled human myotube extract was successfully used as a surrogate to profile the proteome of human muscle tissue [34]. However, this strategy still suffers from the lack of perfect match between the proteome of cultured SILAC myotubes and the proteome of muscle tissue, resulting in several orphan proteins without an internal standard for quantification. Indeed, comparison of the whole proteome of human muscle tissue and cultured human myotubes in our laboratory revealed that 38% of the total identified proteins in muscle tissue extract were not detected in cultured human myotube extract (Fig. 8.5). This could be explained by the fact that muscle tissue is heterogeneous and comprised of a complex mixture of cells besides the muscle fibers, while cultured human myotubes are more homogenous. For example, several components of the neuromuscular junctions, endothelial cells, and conjunctive tissue were detected in muscle tissue but not in cultured human myotubes. Nevertheless, the use of SILAC labeled human myotubes as a surrogate can still be used to quantify a large number of key proteins in human muscle tissue.

A third limitation of super SILAC/SILAM strategy is leaky or missing data when dealing with the analysis of several samples for comparison. This missing measurable data is most often due to lack of MS/MS events and fragmentation of low-abundant peptides in the LC-MS/MS run in data-dependent acquisition

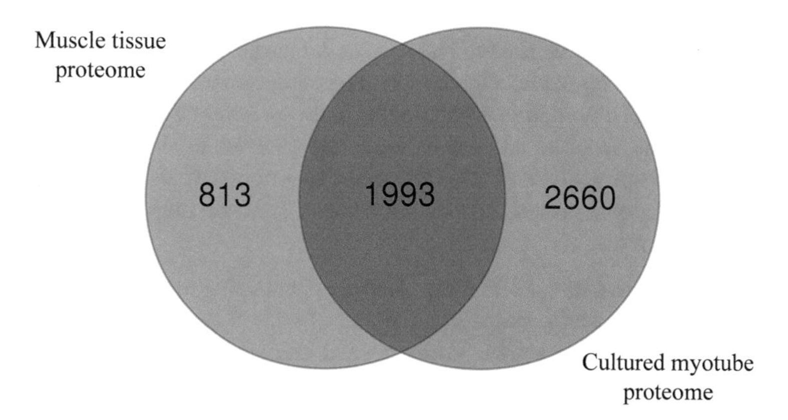

Fig. 8.5 Venn diagram depicting the number of overlapping and unique proteins identified by LC-MS/MS in human muscle tissue and cultured human myotubes. More proteins were identified in cultured human myotubes than in muscle tissue for several reasons. Cultured cells had a lower dynamic range than muscle tissue. Proteins are more easily extracted from cultured cells than from muscle tissue

(DDA) mode. Although the peptide is detected at the MS levels, there is no MS/MS data attached to it. To overcome this issue, a more sophisticated mass spectrometry method such as the sequential window acquisition of all theoretical mass spectra (SWATH-MS) analysis was introduced [35] and is being implemented in several proteomics studies and could be extended to muscle cells or muscle tissue.

8.4 Vital Future Directions

While metabolic labeling by stable isotope amino acids enables accurate quantitative proteomics in muscle cells, muscle tissues, and other tissues, there are still knowledge gaps that may require a combination of the SILAC/SILAM strategies with other additional proteomics methods.

Proteomic Data Interpretation in the Context of Muscle Tissue in Health and Disease Conditions Indeed, most of the time proteomic data is interpreted using knowledge databases that are gathered from proteomics data and literature generated on other tissues and disease conditions which often creates networks that are biased toward cancer and other prevalent diseases. Although muscle tissue expresses about 50% of the total human genome, only a handful of genes (~14%) were defined as being enriched in muscle using a combination of transcriptome data and antibody-based profiling [36]. While this is a great initiative, identifying true muscle-specific proteins using large-scale skeletal muscle proteome profiling will bring insight into the muscle proteome. The next step ideally is to determine the absolute amount of each mass spectrometry identifiable protein in muscle tissue. For example, several key proteins such as dystrophin, utrophin, dysferlin, and dystroglycan are all associated with certain muscle diseases, and knowing the precise levels of these proteins in skeletal muscle under healthy and disease conditions might help define therapeutic targets for gene replacement therapies. SILAC or SILAM strategy in combination with targeted mass spectrometry analysis can easily help determine the absolute amount of these key proteins in skeletal muscle. The next step is perhaps to extend targeted mass spectrometry analyses to absolute quantification of global proteins in skeletal muscle. This will help refine the skeletal muscle proteome atlas.

Protein-Protein Interaction in Muscle Another challenging aspect of working with the muscle proteome, and perhaps in a number of other tissues, is defining protein-protein interaction and networks. While several cell signaling studies and protein interaction studies were facilitated by the use of SILAC and/or SILAM strategies, in other cell models and tissues, only a few studies have been undertaken using muscle cells and muscle tissue. Such studies will bring insight to molecular mechanisms involved in a number of muscle diseases and might help identify novel and innovative therapeutic targets.

Quantitative phosphoproteomics using SILAC and SILAM strategy is another area of interest that could also bring exciting insights about the muscle biology in general but also about alterations in muscle cell signaling under disease conditions.

Qualitative and Quantitative Glycoproteomics of Muscle Tissue Glycoproteins play a crucial role in cell function and tissue integrity. Unfortunately, only a few glycoproteomics studies have been undertaken in skeletal muscle, and most of these studies were focused on the dystrophin-associated protein complex that plays a crucial role in muscle fiber integrity and function. Aberrant assembly and/or glycosylation pattern of this complex has been implicated in a number of muscle diseases and has been a target for thorough glycosylation studies [37]. Extending these analyses to other muscle glycoproteins in combination with SILAC strategy using stable isotope-labeled glutamic acid and SILAM strategy using ^{15}N might help understand the role of glycoproteins in the pathogenesis of dystroglycanopathies [38].

References

1. Gonzalez-Freire, M., Semba, R., Ubaida-Mohien, C., et al. (2016). The human skeletal muscle proteome project: A reappraisal of the current literature. *Journal of Cachexia, Sarcopenia and Muscle, 8*, 5–18. https://doi.org/10.1002/jcsm.12121.
2. Bantscheff, M., Schirle, M., Sweetman, G., et al. (2007). Quantitative mass spectrometry in proteomics: A critical review. *Analytical and Bioanalytical Chemistry, 389*, 1017–1031. https://doi.org/10.1007/s00216-007-1486-6.
3. Ong, S. E., Blagoev, B., Kratchmarova, I., et al. (2002). Stable isotope labeling by amino acids in cell culture, SILAC, as a simple and accurate approach to expression proteomics. *Molecular and Cellular Proteomics, 1*(5), 376–386.
4. Ong, S. E. (2012). The expanding field of SILAC. *Analytical and Bioanalytical Chemistry, 404*(4), 967–976. https://doi.org/10.1007/s00216-012-5998-3.
5. Zhang, A., Williamson, C. D., Wong, D. S., et al. (2011). Quantitative proteomic analyses of human cytomegalovirus-induced restructuring of endoplasmic reticulum-mitochondrial contacts at late times of infection. *Molecular and Cellular Proteomics, 10*(10), M111.009936. https://doi.org/10.1074/mcp.M111.009936.
6. Mintz, M., Vanderver, A., Brown, K. J., et al. (2008). Time series proteome profiling to study endoplasmic reticulum stress response. *Journal of Proteome Research, 7*(6), 2435–2444. https://doi.org/10.1021/pr700842m.
7. Hilger, M., & Mann, M. (2012). Triple SILAC to determine stimulus specific interactions in the Wnt pathway. *Journal of Proteome Research, 11*(2), 982–994. https://doi.org/10.1021/pr200740a.
8. Blagoev, B., Kratchmarova, I., Ong, S. E., et al. (2003). A proteomics strategy to elucidate functional protein-protein interactions applied to EGF signaling. *Nature Biotechnology, 21*(3), 315–318.
9. Mann, M. (2006). Functional and quantitative proteomics using SILAC. *Nature Reviews Molecular Cell Biology, 7*(12), 952–958. https://doi.org/10.1038/nrm2067.
10. Pimienta, G., Chaerkady, R., & Pandey, A. (2009). SILAC for global phosphoproteomic analysis. *Methods in Molecular Biology, 527*, 107–116. https://doi.org/10.1007/978-1-60327-834-8_9.

11. Krüger, M., Moser, M., Ussar, S., et al. (2008). SILAC mouse for quantitative proteomics uncovers kindlin-3 as an essential factor for red blood cell function. *Cell, 134*(2), 353–364. https://doi.org/10.1016/j.cell.2008.05.033.
12. McClatchy, D. B., & Yates, J. R. (2014). Stable isotope labeling in mammals (SILAM). *Methods in Molecular Biology, 1156*, 133–146. https://doi.org/10.1007/978-1-4939-0685-7_8.
13. Zanivan, S., Krueger, M., & Mann, M. (2012). In vivo quantitative proteomics: The SILAC mouse. *Methods in Molecular Biology, 757*, 435–450. https://doi.org/10.1007/978-1-61779-166-6_25.
14. Zanivan, S., Meves, A., Behrendt, K., et al. (2013). In vivo SILAC-based proteomics reveals phosphoproteome changes during mouse skin carcinogenesis. *Cell Reports, 3*(2), 552–566. https://doi.org/10.1016/j.celrep.2013.01.003.
15. Moody, L. R., Barrett-Wilt, G. A., Sussman, M. R., et al. (2017). Glial fibrillary acidic protein exhibits altered turnover kinetics in a mouse model of Alexander disease. *The Journal of Biological Chemistry, 292*(14), 5814–5824. https://doi.org/10.1074/jbc.M116.772020.
16. McClatchy, D. B., Liao, L., Lee, J. H., et al. (2012). Dynamics of subcellular proteomes during brain development. *Journal of Proteome Research, 11*(4), 2467–2479. https://doi.org/10.1021/pr201176v.
17. McClatchy, D. B., Liao, L., Park, S. K., Venable, J. D., & Yates, J. R. (2007). Quantification of the synaptosomal proteome of the rat cerebellum during post-natal development. *Genome Research, 17*(9), 1378–1388.
18. Savas, J. N., Park, S. K., & Yates, J. R. (2016). Proteomic analysis of protein turnover by metabolic whole rodent pulse-chase isotopic labeling and shotgun mass spectrometry analysis. *Methods in Molecular Biology, 1410*, 293–304. https://doi.org/10.1007/978-1-4939-3524-6_18.
19. Le Bihan, M. C., Barrio-Hernandez, I., Mortensen, T. P., et al. (2015). Cellular proteome dynamics during differentiation of human primary myoblasts. *Journal of Proteome Research, 14*(8), 3348–3361. https://doi.org/10.1021/acs.jproteome.5b00397.
20. Giudice, J., & Taylor, J. M. (2017). Muscle as a paracrine and endocrine organ. *Current Opinion in Pharmacology, 34*, 49–55. https://doi.org/10.1016/j.coph.2017.05.005.
21. Grube, L., Dellen, R., Kruse, F., et al. (2018). Mining the secretome of C2C12 muscle cells: Data dependent experimental approach to analyze protein secretion using label-free quantification and peptide based analysis. *Journal of Proteome Research, 17*(2), 879–890. https://doi.org/10.1021/acs.jproteome.7b00684.
22. Hartwig, S., Raschke, S., Knebel, B., et al. (2014). Secretome profiling of primary human skeletal muscle cells. *Biochimica et Biophysica Acta, 1844*(5), 1011–1017. https://doi.org/10.1016/j.bbapap.2013.08.004.
23. Le Bihan, M. C., Bigot, A., Jensen, S. S., et al. (2012). In-depth analysis of the secretome identifies three major independent secretory pathways in differentiating human myoblasts. *Journal of Proteomics, 77*, 344–356. https://doi.org/10.1016/j.jprot.2012.09.008.
24. Duguez, S., Duddy, W., Johnston, H., et al. (2013). Dystrophin deficiency leads to disturbance of LAMP1-vesicle-associated protein secretion. *Cellular and Molecular Life Sciences, 70*(12), 2159–2174. https://doi.org/10.1007/s00018-012-1248-2.
25. Rayavarapu, S., Coley, W., Cakir, E., et al. (2013). Identification of disease specific pathways using in vivo SILAC proteomics in dystrophin deficient mdx mouse. *Molecular and Cellular Proteomics, 12*(5), 1061–1073. https://doi.org/10.1074/mcp.M112.023127.
26. Vila, M. C., Rayavarapu, S., Hogarth, M. W., et al. (2017). Mitochondria mediate cell membrane repair and contribute to Duchenne muscular dystrophy. *Cell Death and Differentiation, 24*(2), 330–342. https://doi.org/10.1038/cdd.2016.127.
27. Vila, M. C., Klimek, M. B., Novak, J. S., Rayavarapu, S., et al. (2015). Elusive sources of variability of dystrophin rescue by exon skipping. *Skeletal Muscle, 5*, 44. https://doi.org/10.1186/s13395-015-0070-6.

28. Shimizu-Motohashi, Y., Komaki, H., Motohashi, N., et al. (2019). Restoring dystrophin expression in duchenne muscular dystrophy: Current status of therapeutic approaches. *Journal of Personalized Medicine, 9*(1), 1. https://doi.org/10.3390/jpm9010001.
29. Anthony, K., Arechavala-Gomeza, V., Taylor, L. E., et al. (2014). Dystrophin quantification: Biological and translational research implications. *Neurology, 83*(22), 2062–2069. https://doi.org/10.1212/WNL.0000000000001025.
30. Wilson, K., Faelan, C., Patterson-Kane, J. C., et al. (2017). Duchenne and becker muscular dystrophies: A review of animal models, clinical end points, and biomarker quantification. *Toxicologic Pathology, 45*(7), 961–976. https://doi.org/10.1177/0192623317734823.
31. Brown, K. J., Marathi, R., Fiorillo, A. A., et al. (2012). Accurate quantitation of dystrophin protein in human skeletal muscle using mass spectrometry. *Journal of Bioanalysis and Biomedicine*, (Supp. 7). https://doi.org/10.4172/1948-593X.S7-001.
32. Hathout, Y., Marathi, R., Rayavarapu, S., et al. (2014). Discovery of serum protein biomarkers in the mdx mouse model and cross-species comparison to Duchenne muscular dystrophy patients. *Human Molecular Genetics, 23*(24), 6458–6469. https://doi.org/10.1093/hmg/ddu366.
33. Geiger, T., Cox, J., Ostasiewicz, P., et al. (2010). Super-SILAC mix for quantitative proteomics of human tumor tissue. *Nature Methods, 7*(5), 383–385. https://doi.org/10.1038/nmeth.1446.
34. Gilmore, J. M., Milloy, J. A., & Gerber, S. A. (2013). SILAC surrogates: Rescue of quantitative information for orphan analytes in spike-in SILAC experiments. *Analytical Chemistry, 85*(22), 10812–10819. https://doi.org/10.1021/ac4021352.
35. Ludwig, C., Gillet, L., Rosenberger, G., et al. (2018). Data-independent acquisition-based SWATH-MS for quantitative proteomics: A tutorial. *Molecular Systems Biology, 14*(8), e8126. https://doi.org/10.15252/msb.20178126.
36. Lindskog, C., Linné, J., Fagerberg, L., et al. (2015). The human cardiac and skeletal muscle proteomes defined by transcriptomics and antibody-based profiling. *BMC Genomics, 16*, 475. https://doi.org/10.1186/s12864-015-1686-y.
37. Townsend, D. (2014). Finding the sweet spot: Assembly and glycosylation of the dystrophin-associated glycoprotein complex. *The Anatomical Record, 297*(9), 1694–1705. https://doi.org/10.1002/ar.22974.
38. Orlando, R. (2012). Quantitative analysis of glycoprotein glycans. *Methods in Molecular Biology, 951*, 197–215. https://doi.org/10.1007/978-1-62703-146-2_13.

Chapter 9
Investigating Muscle Protein Turnover on a Protein-by-Protein Basis Using Dynamic Proteome Profiling

Jatin G. Burniston

9.1 Theory and History of the Technique

Proteomic investigations aim to achieve broad-scale characterisation of the protein complement of muscle and also perform non-targeted differential analysis of the muscle proteome under different conditions (e.g. health versus disease). In the majority, proteomic studies have generated new insight by linking patterns of protein abundance or post-translational modification with different functional states. Such information is regarded as being static because the measurements of abundance or post-translational state represent a 'snapshot' of the muscle proteome under certain conditions at a particular point in time. As such, these data do not include kinetic information and cannot be used to study dynamic aspects of the muscle proteome, including protein turnover or the relative contributions that synthesis and degradation make to changes in protein abundance. For instance, a series of samples collected over time can be used to build a picture of temporal changes in muscle protein abundance, but the question of how the time-dependent changes in the abundance of proteins occurred cannot be answered without also knowing whether (1) the change in a protein's abundance was matched by a greater or lesser rate of synthesis of that protein, and/or (2) whether a change in degradation rate might also have contributed to the difference in protein abundance. Dynamic Proteome Profiling is a new technique that aims to address these questions by offering insight to the synthesis, abundance and degradation of individual proteins in the muscle of humans [1], as well as non-human laboratory animals and cell cultures. Dynamic Proteome Profiling is built on the culmination of a long history of research and methodological development in the fields of stable isotopic labelling, proteomics and computational biology. This chapter aims to highlight the contributions from these separate pillars

J. G. Burniston (✉)
Research Institute for Sport and Exercise Sciences, Liverpool John Moores University, Liverpool, UK
e-mail: j.burniston@ljmu.ac.uk

J. G. Burniston, Y.-W. Chen (eds.), *Omics Approaches to Understanding Muscle Biology*, Methods in Physiology, https://doi.org/10.1007/978-1-4939-9802-9_9

of research and explain how they are brought together in order to perform Dynamic Proteome Profiling in humans.

Stable isotopes are fundamental to investigating the dynamic aspects of the proteome in vivo because they enable proteins to be labelled in a manner that can be detected by mass spectrometry and that does not alter the biological properties of the protein, muscle or organism under study. Rudolf Schoenheimer and co-workers pioneered the application of isotopic tracers in metabolic research in the mid- to late 1930s and provided empirical evidence for the dynamic state of body constituents [2]. Schoenheimer's laboratory developed stable isotope-labelled amino acids [3] and used tyrosine labelled with heavy nitrogen (^{15}N) to generate the first evidence that the amino acids of body proteins are replaced by amino acids from the diet [4]. These findings paved the way for a rich field of research employing isotopic labels to investigate protein and amino acid metabolism. However, isotopic labelling of a particular amino acid (e.g. ^{15}N-lysine) is suboptimal for the technique of Dynamic Proteome Profiling. Labelling just one type of amino acid places a limitation on the relative amount of isotope that can be incorporated into each protein and makes detection and quantitation of isotope incorporation more challenging. Instead, Dynamic Proteome Profiling relies on isotope enrichment through the consumption of 'heavy water'/deuterium oxide (^2H_2O or D_2O) which becomes incorporated into the majority of amino acids during their de novo synthesis or transamination reactions and, therefore, gives a proportionally greater signal for any given rate of protein synthesis (described in more detail later).

Deuterium (^2H or D) is the heavy stable isotope of hydrogen (^1H, 'protium') and was one of the first stable isotopes to be isolated (discovered in 1932 by Harold Urey [5]) and used in biological research. August Krogh was amongst the first to report that ^2H became enriched in the muscle and other tissues of animals that had consumed heavy water. The incorporation of ^2H into muscle was relatively high compared to other tissues and increased yet further after muscle contractions [6, 7]. In 1941, Using [8] used heavy water to report relative rates of deuterium incorporation in protein in vivo. On the basis of separate measurements of whole muscle and the myosin fraction, he deduced that individual proteins within a tissue are likely to have different rates of turnover. Dynamic Proteome Profiling now provides a tool to investigate in detail the range of diversity in the turnover rates of muscle proteins, but for the remainder of the twentieth century, the use of deuterium oxide and the pursuit of individual protein turnover data in vivo were largely unreported. Instead, methods and detection techniques for isotope-labelled amino acids flourished and became the preferred tools for investigating the turnover rates of muscle proteins. For example, Halliday and McKeran [9] was amongst the first report in humans and used infusion of ^{15}N-lysine to investigate whole-body and muscle-specific protein turnover. Consistent with findings from rodent muscle, Halliday and McKeran [9] reported a difference in average synthesis rate of sarcoplasmic (3.8%/d) or myofibrillar (1.5%/d) protein mixtures in human muscle, which is likely underpinned by the different protein compositions of these subcellular fractions.

During the latter part of the twentieth century, the use of stable isotope-labelled amino acids became a well-established technique for investigating human protein and amino acid metabolism (reviewed in [10]). The calculation of the fraction of newly synthesised muscle protein was achieved by measuring the precursor/product ratio after a short period of infusion of a stable isotope-labelled amino acid (described in [11]). Gas chromatography-mass spectrometry (GC-MS) was used to measure the relative abundances of unlabelled 'light' and isotope-labelled 'heavy' amino acids in either muscle protein hydrolysates (product) or the free amino acid (precursor) pools. However, the hydrolysis of muscle proteins to individual amino acid residues severs the link between individual proteins and their synthesis (isotope incorporation) measurement. At the amino acid level, it is no longer possible to determine the protein from which the amino acid originated. As such, the majority of data report fractional synthesis rates that are essentially the gross average of the hundreds or thousands of proteins in hydrolysates of whole muscle or subcellular fractions such as myofibrillar, mitochondrial and sarcoplasmic. This is a significant shortcoming because it is now known that the rate of turnover of individual proteins spans a broad range and there are selective changes to the turnover of individual proteins during physiological and pathophysiological adaptations.

Limited progress was made in the analysis of individual proteins until proteomic techniques became established at around the turn of the twenty-first century. In particular, two-dimensional gel electrophoresis (2DGE) evolved into a robust method for separating proteins based on their isoelectric point and relative mass. This development in protein separation technology co-occurred with the generation of publicly available genome databases and key advancements in mass spectrometry, including new 'soft ionisation' techniques known as electrospray ionisation (ESI) and matrix-assisted laser desorption ionisation (MALDI). The new ionisation techniques meant that large biomolecules such as proteins and peptides could be mass analysed for the first time and earnt their inventors, John Bennett Fenn and Koichi Tanaka, the 2002 Nobel Prize in Chemistry. Routine workflows became established that used 2DGE to separate proteins into individual gel spots that could then be digested with trypsin to produce peptide mixtures for analysis by mass spectrometry. Trypsin digestion of mammalian proteins produces peptides that are typically ~6–20 amino acids long and have masses in the range (~1000–3000 Da) that could now be mass analysed at a sufficient level of resolution to identify the parent protein by comparing the mass spectra against theoretical masses generated by equivalent in silico processing of genome and protein databases.

The introduction of 2DGE and peptide mass spectrometry facilitated a new era of non-targeted 'omic' research at the protein level. Researchers used these techniques to investigate muscle responses to key aspects of physiology such as exercise [12, 13] and ageing [14, 15] by recording 'static' proteome data on the abundance of proteins or their different post-translational states. The dynamic aspects of the muscle proteome were also investigated; for example, Jaleel et al. [16] reports 2DGE separation of proteins extracted from rat gastrocnemius that had been labelled by an infusion of ring-[$^{13}C_6$] phenylalanine in vivo. Two separate but parallel workflows were then applied to (1) identify individual proteins using peptide mass spectrometry and (2) measure the incorporation of isotopically labelled phenylalanine in protein hydrolysates using gas chromatography-mass spectrometry (GC-MS) of derivatised amino acids. These parallel workflows enabled the identity and fractional synthesis rate of 68 mitochondrial proteins to be reported, but this was a relatively laborious solution to acquiring data on protein-specific turnover rates and was not widely adopted in the field. Instead, the ability to mass analyse peptides brought forth new opportunities for investigating protein turnover based solely on peptide mass spectrometry data (providing that sufficient levels of isotope enrichment can be achieved). When peptides are mass analysed, rather than isolated amino acids, it is possible to simultaneously acquire information on the three key aspects that are needed to calculate protein-specific turnover rates: (1) the identity of the parent protein, (2) the level of incorporation of isotopically labelled amino acid in the protein, and (3) the relative enrichment of isotopically labelled amino acid in the precursor pool.

Mass spectrometry resolves peptides as 'envelopes' of mass isotopomers, and the relative abundance of the mass isotopomers reflects the natural abundance of C, H, N and O isotopes in the amino acids that were used to synthesise that protein (Fig. 9.1). Marc Hellerstein and colleagues [17, 18] established an algorithm for mass isotopomer distribution analysis (MIDA) that accurately predicts the relative abundance of m_0, m_1, m_2, m_3, etc., isotopomers using binomial or multinomial expansion of the probabilities of incorporating heavy isotopes based on the size of the molecule and the natural abundance of its elements. Early work [19] used MIDA to measure the synthesis of individual high-abundance proteins (e.g. serum albumin) but used isotope-labelled amino acids at very high levels of precursor enrichment. This approach is impractical for human studies that rely on short (typically <12 h) periods of intravenous infusion, which means the amount of tracer incorporated into new protein is low compared to the large quantity of existing material.

Recently there has been a renewed interest in the use of deuterium oxide (^2H_2O or D_2O, 'heavy water') as a reagent for investigating protein turnover in humans [20]. ^2H_2O has the fundamental advantage that it can be administered via the drinking water so labelling can take place in free-living participants over experimental periods spanning days or months, rather than hours. ^2H_2O is non-hazardous, but it can have isotope effects, including changes to the conformation and stability of biopolymers [21] at concentrations greater than ~20% [22]). This, however, is an order of magnitude greater than the concentrations employed in biosynthetic labelling studies in vivo. For example, ^2H_2O administration in rodents is often designed to achieve a level of 4% enrichment of the body water pool (e.g. [23]). The target level of enrichment is often less than 2% in humans (e.g. [20]), in part, because the cost of achieving high levels of enrichment in humans is prohibitive. At such levels (2–4% MPE), there are no reported long-term adverse effects of ^2H consumption in humans, but acute side-effects (nausea and vertigo) can occur at the beginning of a dosing regimen if ^2H_2O is consumed as a large (i.e. >150 ml) bolus [24]. Therefore, it has become common practice to administer smaller (e.g. 50 ml) doses interspersed throughout the day, and over several days, the deuterium enrichment of the body water compartment rises gradually (Fig. 9.2).

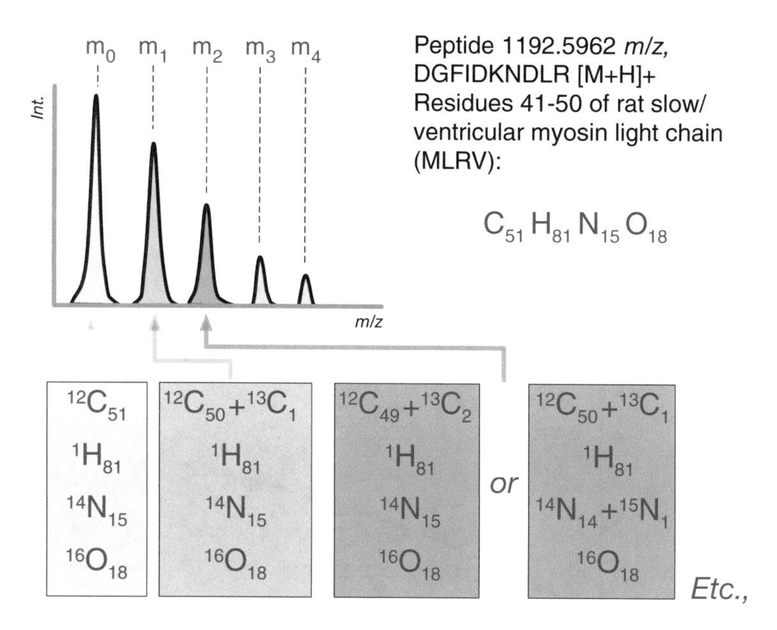

Fig. 9.1 Nomenclature and composition of peptide mass isotopomers. Mass spectrometry resolves peptides as 'envelopes' of mass isotopomers, and the relative abundance of the mass isotopomers reflects the natural abundance of C, H, N and O isotopes in the amino acids that were used to synthesise that protein. The first isotopomer in the series has the lowest mass (mass to charge ratio, *m*/*z*) and is known as the monoisotopic peak (m_0) because it consists entirely of primary/'light' isotopes (i.e. ^{12}C, ^{1}H, ^{14}N, ^{16}O, etc.). The second isotopomer is composed of peptides that contain one stable secondary/'heavy' isotope (i.e. ^{13}C, ^{2}H, ^{15}N, etc.) and is labelled the m_1 peak, and the next isotopomer is labelled the m_2 peak because it contains two secondary/'heavy' isotopes, which may be ^{13}C or other heavy isotopes such as ^{15}N. A peptide with the sequence DGFIDKNDLR has an elemental composition of $C_{51}H_{81}N_{15}O_{18}$. In the absence of an exogenously applied isotope label, the natural pattern of the mass isotopomers is largely determined by the natural abundance of carbon isotopes. This is because carbon is a major component of amino acids and the 'heavy' ^{13}C isotope has a relatively high natural abundance (~1.1% of C is ^{13}C and 98.9% is ^{12}C). The mass isotopomer pattern of a peptide can be roughly predicted based on the probability that 1.1% of the C will be ^{13}C and therefore cannot contribute to the abundance of the m_0 peak. N.B., accurate and complete prediction of the mass isotopomer pattern also requires the contributions of H, N and O isotopes to be recognised and the diminishing probability of two, three or four heavy isotopes occurring with a peptide

$^{2}H_2O$ taken in the drinking water rapidly dissociates to ^{2}HHO and equilibrates with body water in less than 3 h in humans [25] or less than 30 min in rats [23]. Because equilibration of $^{2}H_2O$ with body water and the exchange of ^{2}H into C–H bonds of amino acids each occur rapidly, the rate of protein turnover is the limiting factor to the incorporation of ^{2}H in to protein. Deuterium labelling of amino acids occurs intracellularly, and after incorporation into newly synthesised protein, the ^{2}H-label is irreversible [23]. ^{2}H enrichment of the amino acid precursor pool is equivalent to the body water enrichment [20] and, therefore, can be measured directly in accessible body fluids including serum and saliva using GC-MS analysis

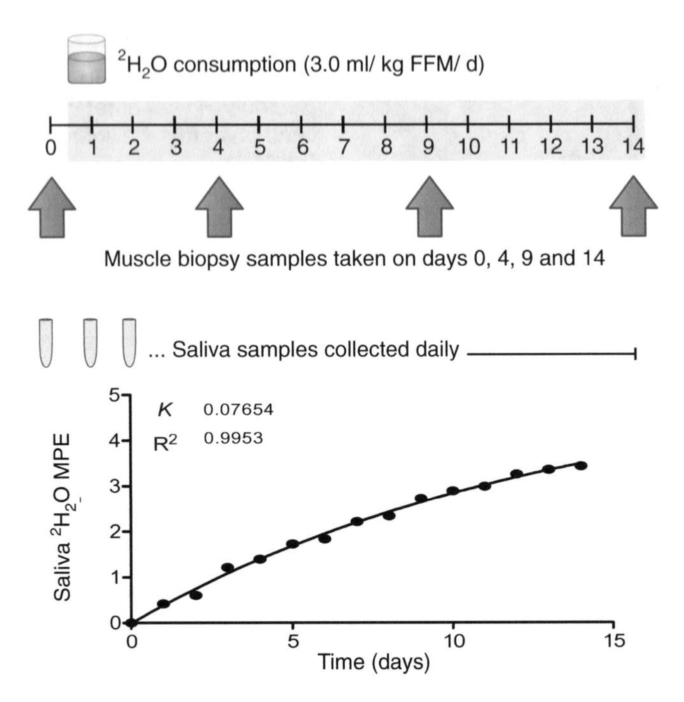

Fig. 9.2 Deuterium oxide/'heavy water' labelling in humans. During an experiment period of 14 days, participants consume 4 × 50 ml 'shots' of deuterium oxide each day and the enrichment of ^2H in their body water compartment rises at a rate (*k*) of 0.07%/d. Percutaneous biopsies of skeletal muscle are taken prior to and at equidistant time points after commencement of ^2H_2O consumption in preparation for Dynamic Proteome Profiling

[26]. These points mean the use of ^2H_2O circumvents several of the technical issues that had complicated the interpretation of data from isotope-labelled amino acid methods. For instance, when using deuterium oxide, the kinetics of precursor enrichment are not confounded by amino acid metabolism or the transport rates and capacities for transport of amino acids across cell membranes. Importantly, deuterium enrichment of the precursor pool is also not diluted by consumption of protein-rich meals or recycling of labelled amino acids due to protein degradation in muscle or other tissues.

Non-essential amino acids such as alanine become labelled with ^2H at their C–H bonds during their de novo synthesis or transamination reactions. Alanine has been a focus of attention because it has four potential sites for labelling and it is one of the most common amino acids (second only to leucine), accounting for ~7.8% of residues in mammalian proteins. In addition, the labelling efficiency of alanine is high (~95% MPE) because of its de novo synthesis from branched-chain amino acids and the transamination of alanine by the reductive amination of pyruvate as part of the alanine-glucose (Cahill) cycle. GC-MS analysis of deuterium-labelled alanine can provide comparable results to amino acid tracer methods in rodents [27] and humans [28, 29] but has the shortcoming that it too severs the link between individual proteins and their synthesis measurement. Highly abundant proteins, such as serum albumin, can be extracted prior to GC-MS analysis (e.g. [20]), but there is a risk of contamination by proteins that may be co-extracted alongside the target of interest.

Mass spectrometry of peptides, rather than isolated amino acids, is not confounded by co-extracted proteins and offers a direct and efficient analytical approach to measuring protein-specific synthesis rates. The majority of amino acids are subject to transamination reactions which involve the exchange of the αC–H and therefore can introduce deuterium ^2H (i.e. αC–^2H) into the intracellular free amino acid pool. The extent of hydrogen exchange in free amino acids in vivo has been empirically determined using tritium in mice [30] and deuterium in humans [31]. After a sustained period of isotope administration, each amino acid reaches an equilibrium that reflects its average level of enrichment from the balance of its transamination reactions, metabolism, and de novo synthesis. In addition to alanine, other non-essential amino acids, such as glycine and glutamine/glutamic acid, also exhibit high levels of deuterium enrichment. As a consequence, deuterium incorporation into protein is sufficient for it to be detected on a protein-by-protein basis using existing proteomic profiling methods.

A common approach for proteome profiling uses 'bottom-up' analysis, which involves digestion of proteins followed by delivery of the peptide mixture into a mass spectrometer by reversed-phase liquid chromatography (Fig. 9.3). The peptide mixture is resolved in time, and a chromatogram is produced wherein the intensity of each peptide peak provides a measure of abundance. Label-free quantitation is performed, which defines the isotopic envelope of each peptide and records the abundance of all mass isotopomers over the duration of the chromatographic peak for that peptide. Several software platforms exist for label-free quantitation of protein abundances, and log-transformed MS data can be exported after normalisation by inter-sample abundance ratio and used to investigate differences in protein abundance [32, 33]. However, there are currently no commercially available software applications for automated analysis of protein synthesis rates, and researchers working in this area generally develop their own solutions for computing synthesis rates from peptide mass isotopomer data.

Similar to the aforementioned application of MIDA, the isotopomer distribution of deuterium-labelled peptides can also be used to calculate the fractional rate of synthesis of a protein [34, 35]. The data processing is complex but not difficult and the fundamentals of the process have been reviewed in detail [36, 37]. The number (n) of exchangeable H atoms must be calculated from the peptide sequence with reference to standard tables reporting the enrichment of amino acids by deuterium [31]. The peptide mass isotopomer distribution of deuterium-labelled peptides is compared to either the theoretically predicted natural distribution of mass isotopomers for that peptide or the empirically measured mass isotopomer distribution of the peptide in unlabelled samples. We use baseline subtraction of the natural mass isotopomer distribution because this reduces error by accounting for differences in instrument performance on a sample-by-sample basis. Subtraction of the natural/unlabelled mass isotopomer distribution from mass spectra of deuterium-

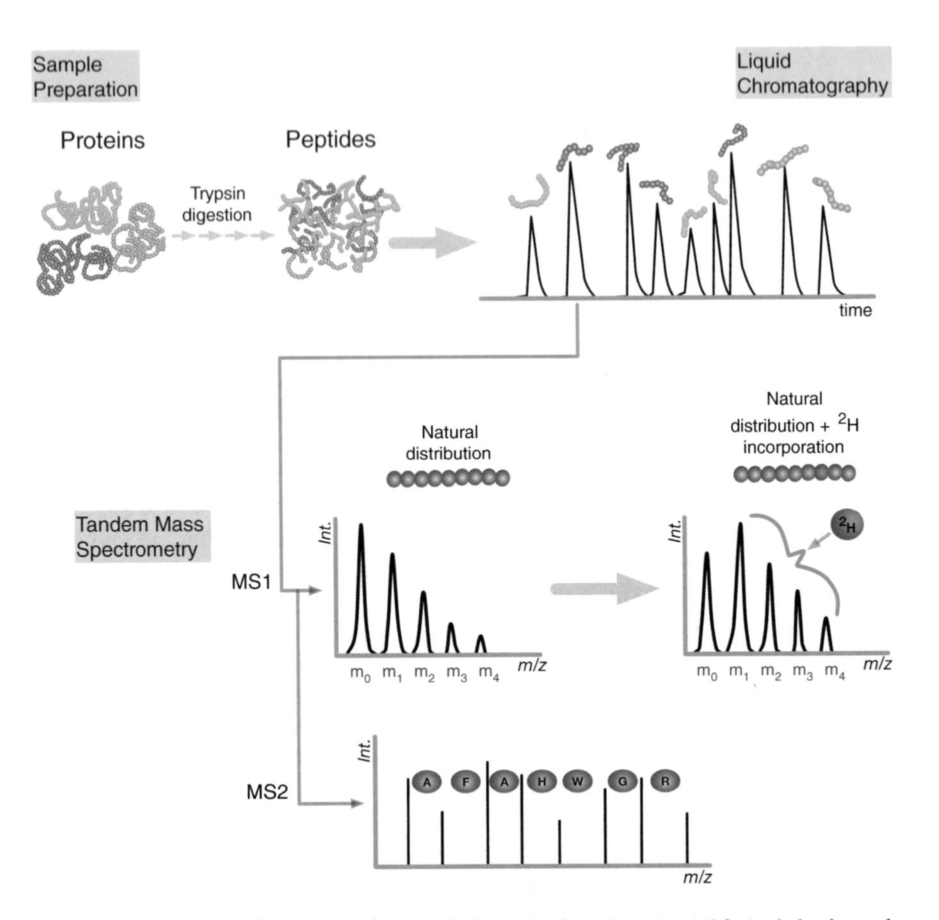

Fig. 9.3 Proteomic profiling of deuterium-labelled muscle. Proteins extracted from skeletal muscle samples taken prior to and after a period of ^2H_2O labelling in vivo are digested into peptides of ~6–20 amino acids (circles). The peptide mixture is separated in time by reversed-phase liquid chromatography. Peptides elute based on their relative hydrophobicity and are delivered to the mass spectrometer. Tandem mass spectrometry (MS/MS) involves two levels and mass analysis: Level 1 (MS1) records the mass spectra of the intact peptide and level 2 (MS2) records the mass spectra of each peptide after it has undergone fragmentation. MS1 data are used to analyse both the relative abundance (based on the intensity of all mass isotopomers) and deuterium (^{2}H) incorporation (based on the relative distribution of mass isotopomers) of each peptide. MS2 data provide information on the peptide sequence (i.e. AFAHWGR) and are used to identify peptides by comparing the pattern of fragmentation against protein databases. ^{2}H-labelled amino acids incorporated into protein during synthesis in vivo cause a shift in the distribution of MS1 peptide mass isotopomers. Peptides that contain ^{2}H-labelled amino acids can only contribute to the abundance of m_1, m_2, m_3, etc. mass isotopomers, and the relative abundance of the m_0 mass isotopomer declines as a function of deuterium incorporation (protein synthesis)

labelled peptide gives a measure of the isotopomer excess (or molar percent excess, MPE) due to the incorporation of deuterium in vivo. The ratio between 'heavy' isotope-containing peaks (e.g. m_2/m_1 peaks) can be used to back-calculate the relative enrichment of the aminoacyl-tRNA precursor pool (p), but more often direct measurements of body water enrichment are used via GC-MS analysis of accessible bodily fluids such as plasma or saliva.

When the level of precursor enrichment is known, the incorporation of deuterium (i.e. fraction of newly synthesised protein) can be calculated from the ratio of unlabelled (m_0)/labelled (m_1) mass isotopomers. Alternatively, the ratio of unlabelled (m_0) can be expressed relative to the sum of labelled (m_1, m_2, m_3, etc.) isotopomers, which gives a proportionally greater magnitude of change for any given amount of ^2H incorporation. The monoisotopic peak (m_0) cannot contain heavy isotopes; therefore the molar fraction of m_0 declines as a function of ^2H incorporation into protein. Hence, while biosynthetic labelling of protein turnover (i.e. incorporation of ^2H in to protein) exhibits asymptotic 'rise-to-plateau kinetics, data analysis is based on the reciprocal decline in the molar fraction of m_0, which follows the pattern of an exponential decay:

$$f_t = f_{t0} \bullet e^{-kt} \quad (9.1)$$

The rate (k) of decay of the molar fraction (f) of the monoisotopic peak can be calculated from the change in m_0 from prior to (t_0) to after (t_1) a defined period of deuterium incorporation using the formula:

$$k = 1/(t - t_0) \bullet \ln(f_t) - \ln(f_{t0}) \quad (9.2)$$

The rate of decay (k) of m_0 must then be normalised by the number (n) of exchangeable H positions based on the peptide sequence and the level of enrichment (p) of the precursor pool (e.g. molar percent enrichment of ^2H in the body water compartment). The above 2-point calculation {i.e. samples taken at baseline (t_0) and after a period of deuterium consumption (t_1)} is sufficient to estimate the fractional synthesis rate of proteins when precursor enrichment can be rapidly raised and then maintained at a steady state (such as for experiments in cell cultures or laboratory rodents). However, these criteria are not readily achieved in humans that consume ^2H_2O via their drinking water, and in humans there is a gradual rise in precursor enrichment even when multiple doses of heavy water are consumed each day (Fig. 9.2).

In humans, the calculation of the rate of ^2H into protein (i.e. synthesis rate) must take into account the rise-to-plateau kinetics of ^2H in the precursor/body water compartment in addition to the kinetics of the incorporation of deuterium into protein. Therefore, ^2H incorporation into protein exhibits bi-exponential kinetics and calculations to estimate protein fractional synthesis rates use approaches that have developed from graphical approaches for the interpolation of data on free versus bound radioisotope-labelled amino acid tracers [38]. Mathematical techniques (e.g. Nelder-Mead optimisation) can be deployed to fitting bi-exponent data on precursor enrichment and the incorporation of ^2H into protein [39] provided that data is available across multiple time points. As a minimum, protein synthesis

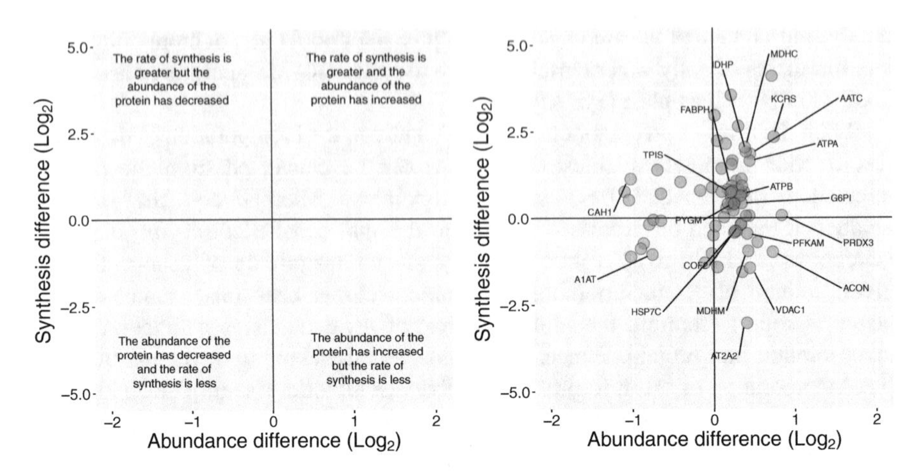

Fig. 9.4 The dynamic proteome response of human muscle to exercise. Dynamic Proteome Profiling data can be presented on quadrant plots to illustrate the relationship between changes in abundance and synthesis rate of individual proteins (left panel). The panel on the right illustrates preliminary unpublished data on the response of human muscle to high-intensity interval training and is annotated with UniProt protein identifiers. Proteins such as ATP synthase subunit alpha (ATPA, top-right quadrant) increase in abundance and are synthesised at a greater rate in exercised muscle, whereas sarcoplasmic/endoplasmic reticulum ATPase 2 (AT2A2, bottom-right quadrant) is more abundant in exercised muscle despite its rate of synthesis being less compared to baseline/pre-exercise levels

calculations in humans require serial analysis encompassing the baseline (i.e. in the absence of deuterium) and at least three subsequent time points after deuterium administration has commenced (Fig. 9.2). This is necessary to capture information on the bi-exponential pattern of change in peptide mass isotopomer distribution.

Dynamic Proteome Profiling brings together protein-specific data on fractional synthesis rates with protein abundance measurements to give unique insight to the contribution of degradation to changes in protein abundance in vivo. The rate of change in the abundance of a protein is dependent on the difference between its rate of synthesis and the rate of degradation. Therefore, by converting differences in abundance between baseline and post-labelling time points into rates of change [i.e. by substituting abundance data into Eq. (9.2)], the rate of degradation of each protein is calculated as the difference between its rate of synthesis and its rate of change in abundance. Furthermore, by plotting changes in abundance with changes in synthesis rate (Fig. 9.4), it is possible to visualize different patterns of response to an experimental intervention [1].

9.2 Major Applications

At present the majority of literature relevant to Dynamic Proteome Profiling in vivo has been generated from the analysis of the brain [40], liver [41] or heart [39, 42] of laboratory rodents. Compared to humans, the body mass of rodents is smaller and the rate of protein turnover is higher; therefore relatively little deuterium oxide is required and the duration of the labelling experiment can be relatively short. In addition, deuterium enrichment of the body water compartment can be raised rapidly and then maintained at a steady state by delivering an intraperitoneal bolus followed by supplementation of the animal's drinking water with ^2H_2O. A constant level of precursor enrichment eases the burden of data analysis because only the kinetic parameters of deuterium incorporation in newly synthesised protein need be considered, i.e. mono-exponential rise-to-plateau or 'single-compartment' models can be used. Lastly, relatively high levels of precursor enrichment (e.g. 4% MPE) can be achieved in rodents without being prohibitively expensive, and large quantities of tissue can be collected for analysis, each of which can help to increase the sensitivity of the analysis.

Kim et al. [42] report one of the earliest large-scale analyses of protein turnover in striated muscle and investigated mouse heart mitochondria using ^2H_2O labelling in vivo and bottom-up proteomic analysis of tryptic peptides. The synthesis rates of 314 cardiac mitochondrial proteins were characterised and the median rate of synthesis was 0.04 d^{-1}, which corresponds to a median half-life of ~18 days. Liver mitochondria were also analysed, and, generally, protein synthesis rates were threefold greater in liver than heart, but there was a broader range (i.e. 75-fold range spanning 0.0112 d^{-1} to 0.8592 d^{-1}) of synthesis rates in cardiac mitochondria compared to the ~35-fold range (0.0211 d^{-1} to 0.7441 d^{-1}) in liver. The correlation between synthesis rates of proteins in heart and liver mitochondria was modest (Spearmen $\rho = 0.50$), and more detailed inspection reveals the synthesis rates of mitochondrial proteins rank differently between the heart and liver, which suggests tissue-specific regulation of protein turnover. We [34] reported similar findings across a selection of rat striated muscles, including heart, diaphragm, slow-twitch soleus and fast-twitch extensor digitorum longus. We found that not only is the synthesis rate of each protein different in each muscle but also the rank-order of synthesis rates was different from muscle to muscle, which suggests the general physiochemical properties of proteins cannot be used to predict their rate of turnover. Indeed, there is no relationship between turnover rate and the protein's molecular weight, isoelectric point, hydrophobicity or motifs such as PEST (proline, glutamate, serine and threonine enriched sequences) that have previously been associated with proteins that have a short half-life [42].

It is also not obvious that protein turnover clusters with protein functional groups. Cardiac mitochondria proteins involved in protein folding and proteolysis are amongst those most rapidly synthesised, whereas the majority of proteins involved in oxidative phosphorylation are amongst those with the lowest rates of synthesis [42]. However, these are broad generalisations and there are some notable exceptions to these trends. For example, the synthesis of NADH dehydrogenase assembly factor 3 (NDUFAF3) was approximately double the average rate of complex I subunit proteins, which may point to a specific regulatory role for this protein in complex I topogenesis and assembly. Proteomic synthesis data also argue against a dominant role for turnover of whole mitochondria, e.g. via mitophagy, in the normal heart because the turnover rate of proteins located on the outer mitochondrial membrane is generally greater than the turnover rate of inner mitochondrial membrane proteins. In the future it will also be important to study the relative abundances of proteins alongside their turnover rates to capture the full picture on proteome dynamics.

Broader analysis is also needed to determine how the turnover of mitochondrial proteins relates to the wider muscle proteome and the discrete subcellular populations of mitochondria in muscle. Mitochondria in the heart, as well as in skeletal muscle, are present as at least two discrete subcellular populations: interfibrillar mitochondria and subsarcolemmal mitochondria [43] that have different functional characteristics and exhibit different responses in models of disease or exercise training [44]. Generally, interfibrillar mitochondria have a higher oxidative capacity and greater responsiveness to exercise, but Kasumov et al. [45] report protein-specific synthesis rates are lower in interfibrillar compared to subsarcolemmal mitochondria. Therefore, comparatively low rates of turnover under basal conditions may not automatically be interpreted to mean a poor responsiveness to external stimuli. Indeed, changes to the rate of synthesis of proteins occur during muscle adaptation; therefore basal rates of protein turnover cannot be used to predict the rate of change (potential for adaptation) of the muscle proteome in response to stimuli.

Research investigating the synthesis of individual proteins in the muscle of humans is in its infancy. The earliest proteome-scale data was published in 2016 [46] and reports a cross-sectional comparison of muscle from sedentary adults and individuals that had undertaken a 3-week regimen of sprint-interval cycling. After stringent filtering of the mass spectrometry data, 139 proteins were analysed in at least $n = 3$ sedentary versus $n = 3$ exercised participants and 20 proteins had significantly greater rates of synthesis in exercised muscle. The data point to a greater synthesis of proteins involved in glucose metabolic processes, apoptosis, contraction and cellular respiration, but further work is needed to specifically compare the effects of exercise against baseline values to rule out potential differences in the abundance of proteins between sedentary and exercised individuals. Improvements in exercise performance after high-intensity aerobic exercise are associated with both increases and decreases in the abundance of select proteins in human muscle [47]. Increases in the abundance of the alpha and beta subunits of the ATP synthase complex are a prominent feature of exercise-trained muscle [13, 48], and Shankaran et al. [46] report the synthesis of these proteins was greater in the muscle of participants that performed sprint-interval training. Shankaran et al. [46] also report exercised muscle has a greater synthesis of glycolytic enzymes, but we [47] found no difference in the abundance of these proteins after high-intensity interval training. This dissimilarity could be due to slight differences in training regimens, but, moving forward, it will be important to investigate which exercise-induced changes in protein synthesis also result in changes in protein abundance using more comprehensive Dynamic Proteome Profiling.

The synthetic response of human muscle to endurance training interventions, such as sprint-interval cycling, has been a largely neglected area of investigation in the earlier literature from amino acid tracer studies. In part, this is because experiments using isotopically labelled amino acids are more suited to capturing responses to acute interventions (e.g. hormones, protein feeding or resistance exercise) that have large and immediate effects on the synthesis of abundant myofibrillar proteins [49]. That said, the large acute synthetic response of muscle in the hours immediately after a single bout of resistance exercise tends to overestimate the cumulative growth that occurs during long-term training programmes [50]. Likewise, dietary interventions designed to optimise the acute response to a single bout of exercise have less obvious benefits when the response is integrated over days or weeks of resistance training. For example, the synthesis of mixed muscle protein in the hours immediately after a bout of resistance exercise is significantly greater when dietary protein consumption is distributed evenly across four meals each day compared to when protein is consumed in a skewed pattern, i.e. when the majority of protein is consumed at dinner [51]. This beneficial effect was absent in follow-up analysis of the experiment that reported time-integrated synthetic responses across protein mixtures using ^2H_2O labelling [28]. Nevertheless, proteomic analysis of ^2H_2O-labelled muscle reported 68 individual proteins that had significantly greater rates of synthesis during the period of resistance training compared to baseline values. Importantly this synthetic response to resistance exercise was evident across different functional classes of proteins that included respiratory enzymes in addition to myofibrillar proteins. The effect of resistance exercise on respiratory enzymes had not been detectible from the analysis of protein mixtures and gives wider insight to the muscle response to resistance training.

In the aforementioned works, the analysis of protein synthesis rates was not routinely matched with equivalent analysis of protein abundances. Therefore, the individual components of protein turnover, i.e. synthesis, abundance and degradation, were not investigated. Herein, we use the term 'Dynamic Proteome Profiling' to describe a more comprehensive technique that gives insight to the individual components of protein turnover by combining protein-specific synthesis measurements with established [32] label-free proteomic profiling methods that can also report the relative abundance of each protein. We [1] have used Dynamic Proteome Profiling to study protein-by-protein responses of human muscle to resistance exercise training. Our analysis included 90 of the most abundant muscle proteins encompassing major myofibrillar components and key metabolic enzymes that represent ~85% of the total protein mass of muscle [52]. To facilitate robust statistical analysis, we filtered out proteins that could not be detected in each of the eight control and eight exercise-trained participants across all four of the sampling points. A further unique feature of our work [1] was that we specifically investigated proteome dynamics in a changing system.

Almost without exception, previous biosynthetic labelling experiments have been designed such that protein abundance does not change during the period of investigation, and, therefore, synthesis rates are assumed to be equivalent to rates of degradation. When both protein abundance and synthesis are measured directly, it is possible to investigate a system that is undergoing change, wherein increases or decreases in protein abundances can be explained in terms of the relative contributions of synthesis and degradation. In response to resistance exercise, human muscle proteins exhibit several different patterns of response [1], including:

1. Enhanced turnover (i.e. increased synthesis rate but no change in protein abundance).
2. Changes in abundance that are primarily driven by degradation (i.e. increases in abundance in the absence of increases in synthesis rate or decreases in abundance despite increases in synthesis).
3. Proteins that increased in abundance because their synthesis rate was elevated.

These findings offer a novel glimpse of the intricacies of muscle adaptation by highlighting roles for protein turnover and selective protein degradation in addition to purely synthetic accretion of protein in the mechanisms underlying changes in the muscle proteome.

9.3 Limitations

Dynamic Proteome Profiling is unique in its ability to provide insight to the individual components of protein turnover (i.e. synthesis, abundance and degradation) on a protein-by-protein basis in humans. The analysis is reproducible [53], but the technique is very much in its infancy so there is undoubtedly scope for further optimisation. The use of heavy water for biosynthetic labelling is less invasive than infusion of isotope-labelled amino acids, and studies can be conducted in free-living participants; nevertheless the participant burden of Dynamic Proteome Profiling studies can be relatively demanding. Participants may be required to consume ^2H_2O several times throughout each day, provide daily saliva samples or frequent venous blood samples and undergo percutaneous needle biopsies in addition to the demands of the physiological intervention. Dynamic Proteome Profiling is best suited to studying long-term integrated protein dynamics over period of days or weeks in humans, but it is not optimal for studying the acute short-term (e.g. <1 d) responses because the equilibration of ^2H in body water is relatively slow and the consumption of large boluses is associated with discomforting side-effects.

Some limitations of Dynamic Proteome Profiling are common to all proteomic endeavours in striated muscle. That is, while skeletal muscle is an accessible and important body tissue, it is a technically challenging substrate for proteomic investigation because it contains a small number of high-abundance proteins that dominate the analytical landscape. Indeed, approximately half of the protein content of muscle is accounted for by just ten proteins, which makes it relatively difficult to achieve deep proteome mining. Furthermore, the broad diversity of muscle phenotypes is underpinned by complex patterns of co-expression of fast and slow isoforms of myofibrillar proteins that share high levels of sequence homology. This makes peptide-level studies particularly challenging because isoform-specific peptides must be detected in order to study adaptations of the myofibrillar proteome.

As is the case with all proteomic analyses, the depth of the analysis (i.e. total number of proteins reported) can be greatly enhanced by fractionating samples and exploiting orthogonal separation techniques prior to analysis using the most up-to-date high-resolution mass spectrometers. However, in any proteomic experiment, the number of proteins that can be identified will be greater than the number that can be quantified because of issues regarding missing data. That is, to identify a protein, it need only be detected in one sample amongst many, but to quantify a protein, peptides that are unique to that protein must be clearly resolved in each biological replicate. Often this issue is partially circumvented by accepting a less than full number of biological replicates for statistical testing of proteomics data, but this is not best practice. Dynamic Proteome Profiling adds a further layer of complexity and requires high-quality mass isotopomer distribution profiles to be captured for each peptide during their entire chromatographic profile. Therefore, some peptides may be excluded because co-eluting peptides with similar or overlapping mass isotopomer envelopes contaminate each other and so the number of proteins submitted to statistical analyses is further reduced.

Because of the above-mentioned quality control processes, the analysis of some proteins is based on data from single peptides, whereas other proteins may have numerous isoform-specific peptides that are well-resolved and submitted to statistical analysis. Profiling of the same peptide across different experimental groups is robust, but it is not yet certain how representative one peptide is of the abundance or synthesis rate of the entire protein. Closer inspection of proteins with numerous quantifiable isoform-specific peptides sometimes reveals that peptides from the same protein exhibit broadly different rates of synthesis. Currently it is routine practice to report the median rate across all peptides belonging to a protein to summarise the average synthesis rate. Given the high level of reproducibility of peptide mass isotopomer data, the differences in peptide synthesis rates are unlikely to be due entirely to technical errors. In the future peptide-specific data could be used to gain insight to domain-specific rates of turnover. For example, they may indicate peptide domains that may be cleaved or subject to post-translational modifications that alter their rate of turnover. Initially, it may be difficult to resolve these issues using bottom-up analysis of peptide mixtures, but top-down analysis of proteins using 2DGE could be used to investigate proteoform-specific synthesis rates and shed new light on whether a particular post-translation modification is associated with a change to the synthesis or degradation rate of that protein.

In order to begin interrogating more complex proteoform-specific responses, it will be necessary to further optimise the experiment designs and mathematical models for generating protein synthesis rates from peptide mass spectrometry data. Experiments that collect only baseline and post-intervention samples are less appropriate for detailed interrogation of protein responses because they do not chart the time course of changes in deuterium incorporation and therefore cannot establish whether labelling of a protein has reached equilibration. Time series analyses (i.e. including $\geq$4 samples) are better suited to detailed studies, but the timing of sample collection dictates the sensitivity of the synthesis measurements, and this design may be less appropriate for clinical settings because the collection of numerous muscle samples may be less tolerated.

9.4 Vital Future Directions

Protein turnover is the fundamental mechanism that both maintains proteome quality and enables muscle adaptation/changes in protein abundance to occur. Over recent decades there has been great focus on the molecular biology of gene transcription and signal transduction pathways (e.g. governing ribosomal translation) as the mechanisms and regulators of protein abundances. It is important to appreciate that the abundance of a protein can also be modulated by changes to its rate of degradation and that changes to the rate of replacement/turnover of proteins may influence muscle function in the absence of changes in gene expression or protein abundance. Indeed, there is a growing awareness of the importance of 'quality control' particularly regarding the biology of ageing [54] and as a mechanism of disease. This is driving the development of methods such as Dynamic Proteome Profiling to investigate protein homeostasis (i.e. proteo-stasis) or protein dynamics (i.e. proteo-dynamics).

Some experimental situations and tissues cannot be studied in humans, e.g. sampling of vital organs is not commonplace, but skeletal muscle is readily accessible using percutaneous biopsy techniques. Omic studies in human muscle provide a rare opportunity for hypothesis generating research into ageing, chronic human diseases and exercise training which may each be underpinned by complex gene-environment interactions that are challenging to model in laboratory rodents. Muscle is the most abundant tissue in healthy adults, and it makes a significant contribution to the health and well-being of individuals of all ages. Little work has been done to investigate the mechanistic role that protein turnover plays in muscle adaptation, but 'quality control' at the protein level may be particularly important in skeletal muscle because turnover at the cellular level involving regeneration and differentiation of myofibres is not considered to be a prominent feature of healthy adult muscle. Therefore, Dynamic Proteome Profiling offers an important new tool for gaining insight to the regulation of human muscle health.

We have studied the dynamic responses of muscle to exercise, and in the future, it will be of interest to know whether diseases, such as type 2 diabetes, that are associated with impaired muscle function also exhibit differences in the quality as well as abundance of muscle proteins. Similarly, losses in muscle mass, whether as a consequence of ageing or due to disuse, injury or disease, will be an important area for the application Dynamic Proteome Profiling. As with all new techniques, there is scope for further enhancement and optimisation of the assay. Deuterium labelling is relatively inexpensive for studies in cell culture and small laboratory animals, but studies in humans currently require large amounts of deuterium and the cost can be prohibitive. Dose-response studies are required to find the limits of sensitivity of peptide mass spectrometry for detecting deuterium incorporation in human muscle proteins. The burgeoning development of mass spectrometry instrumentation will further impact on the minimum level of precursor enrichment required to perform Dynamic Proteome Profiling and will also enable a greater proportion of the muscle proteome to be interrogated. That said, user-friendly tools for data analysis and bioinformatics are currently the major obstacle to widespread deployment of Dynamic Proteome Profiling in physiology. Generally, bioinformatic tools for proteomics are more complex and lag behind other omic disciplines, which is unfortunate because the proteome is the most direct molecular link to cell function and has the greatest potential to bring new insight in physiology. With regard to Dynamic Proteome Profiling, developments are required to improve the robustness of curve-fitting processes and to streamline the downstream bioinformatic processing of data that encompasses synthesis, abundance and degradation values on a proteome-wide level.

Synthesis and degradation are regulated independently and on a protein-by-protein basis to affect the abundance and half-life of each protein. Presently, little is known regarding the turnover of individual proteins, particularly in human muscle, and the role that protein degradation plays as a regulatory mechanism in muscle adaptation is largely unexplored. Early findings from Dynamic Proteome Profiling of the human muscle response to resistance exercise [1] highlight that muscle adaptation involves a range of different dynamic responses, including instances where degradation has the dominant influence on changes in protein abundance. Generally, the magnitude of the dynamic response is greater than the magnitude of change in protein abundance, indicating that kinetic measurements of protein turnover may be a more sensitive marker of muscle response than changes in protein abundance. In addition, it is clear that changes to the turnover/replacement rate of proteins are a response in and of itself. Therefore, it is key that simultaneous measurements of synthesis and abundance are made in order to distinguish changes to the rate of protein replacement from effects that lead to a change in protein abundance.

References

1. Camera, D. M., Burniston, J. G., Pogson, M. A., Smiles, W. J., & Hawley, J. A. (2017). Dynamic proteome profiling of individual proteins in human skeletal muscle after a high-fat diet and resistance exercise. *The FASEB Journal, 31*, 5478–5494.
2. Schoenheimer, R., & Clarke, H. T. (1942). *The dynamic state of body constituents*. Cambridge, MA: Harvard University Press.
3. Schoenheimer, R., & Ratner, S. (1939). Studies in protein metabolism: III. Synthesis of amino acids containing isotopic nitrogen. *Journal of Biological Chemistry, 127*, 301–313.

4. Schoenheimer, R., Ratner, S., & Rittenberg, D. (1939). The process of continuous deamination and reamination of amino acids in the proteins of normal animals. *Science, 89*, 272–273.
5. Urey, H., Brickwedde, F., & Murphy, G. (1932). A hydrogen isotope of mass 2. *Physical Review, 39*, 164–165.
6. Krogh, A., & Ussing, H. H. (1936). The exchange of hydrogen between the free water and the organic substances in the living organism. *Acta Physiologica, 75*, 90–104.
7. Ussing, H. H. (1937). The exchange of H and D atoms between water and protein in vivo and in vitro. *Acta Physiologica, 77*, 107–122.
8. Ussing, H. H. (1941). The rate of protein renewal in mice and rats studied by means of heavy water. *Acta Physiologica Scandinavica, 2*, 209–221.
9. Halliday, D., & McKeran, R. O. (1975). Measurement of muscle protein synthetic rate from serial muscle biopsies and total body protein turnover in man by continuous intravenous infusion of L-(alpha-15N)lysine. *Clinical Science and Molecular Medicine, 49*, 581–590.
10. Wagenmakers, A. J. (1999). Tracers to investigate protein and amino acid metabolism in human subjects. *The Proceedings of the Nutrition Society, 58*, 987–1000.
11. Kim, I. Y., Suh, S. H., Lee, I. K., & Wolfe, R. R. (2016). Applications of stable, nonradioactive isotope tracers in in vivo human metabolic research. *Experimental & Molecular Medicine, 48*, e203.
12. Burniston, J. G. (2008). Changes in the rat skeletal muscle proteome induced by moderate-intensity endurance exercise. *Biochimica et Biophysica Acta, 1784*, 1077–1086.
13. Burniston, J. G., & Hoffman, E. P. (2011). Proteomic responses of skeletal and cardiac muscle to exercise. *Expert Review of Proteomics, 8*, 361–377.
14. Doran, P., Donoghue, P., O'Connell, K., Gannon, J., & Ohlendieck, K. (2009). Proteomics of skeletal muscle aging. *Proteomics, 9*, 989–1003.
15. Gelfi, C., Vigano, A., Ripamonti, M., Pontoglio, A., Begum, S., Pellegrino, M. A., Grassi, B., Bottinelli, R., Wait, R., & Cerretelli, P. (2006). The human muscle proteome in aging. *Journal of Proteome Research, 5*, 1344–1353.
16. Jaleel, A., Short, K. R., Asmann, Y. W., Klaus, K. A., Morse, D. M., Ford, G. C., & Nair, K. S. (2008). In vivo measurement of synthesis rate of individual skeletal muscle mitochondrial proteins. *American Journal of Physiology Endocrinology and Metabolism, 295*, E1255–E1268.
17. Hellerstein, M. K., & Neese, R. A. (1999). Mass isotopomer distribution analysis at eight years: Theoretical, analytic, and experimental considerations. *The American Journal of Physiology, 276*, E1146–E1170.
18. Hellerstein, M. K., & Neese, R. A. (1992). Mass isotopomer distribution analysis: A technique for measuring biosynthesis and turnover of polymers. *The American Journal of Physiology, 263*, E988–E1001.
19. Papageorgopoulos, C., Caldwell, K., Shackleton, C., Schweingrubber, H., & Hellerstein, M. K. (1999). Measuring protein synthesis by mass isotopomer distribution analysis (MIDA). *Analytical Biochemistry, 267*, 1–16.
20. Previs, S. F., Fatica, R., Chandramouli, V., Alexander, J. C., Brunengraber, H., & Landau, B. R. (2004). Quantifying rates of protein synthesis in humans by use of $2H_2O$: Application to patients with end-stage renal disease. *American Journal of Physiology Endocrinology and Metabolism, 286*, E665–E672.
21. Thomson, J. F. (1960). Physiological effects of D20 in mammals. *Annals of the New York Academy of Sciences, 84*, 736–744.
22. Barbour, H. G. (1937). The basis of the pharmacological action of heavy water in mammals. *The Yale Journal of Biology and Medicine, 9*, 551–565.
23. Busch, R., Kim, Y. K., Neese, R. A., Schade-Serin, V., Collins, M., Awada, M., Gardner, J. L., Beysen, C., Marino, M. E., Misell, L. M., & Hellerstein, M. K. (2006). Measurement of protein turnover rates by heavy water labeling of nonessential amino acids. *Biochimica et Biophysica Acta, 1760*, 730–744.
24. Jones, P. J., & Leatherdale, S. T. (1991). Stable isotopes in clinical research: Safety reaffirmed. *Clinical Science (London, England), 80*, 277–280.

25. Holm, L., O'Rourke, B., Ebenstein, D., Toth, M. J., Bechshoeft, R., Holstein-Rathlou, N. H., Kjaer, M., & Matthews, D. E. (2013). Determination of steady-state protein breakdown rate in vivo by the disappearance of protein-bound tracer-labeled amino acids: A method applicable in humans. *American Journal of Physiology Endocrinology and Metabolism, 304*, E895–E907.
26. McCabe, B. J., Bederman, I. R., Croniger, C., Millward, C., Norment, C., & Previs, S. F. (2006). Reproducibility of gas chromatography-mass spectrometry measurements of 2H labeling of water: Application for measuring body composition in mice. *Analytical Biochemistry, 350*, 171–176.
27. Gasier, H. G., Riechman, S. E., Wiggs, M. P., Previs, S. F., & Fluckey, J. D. (2009). A comparison of $2H_2O$ and phenylalanine flooding dose to investigate muscle protein synthesis with acute exercise in rats. *American Journal of Physiology Endocrinology and Metabolism, 297*, E252–E259.
28. Murphy, C. H., Shankaran, M., Churchward-Venne, T. A., Mitchell, C. J., Kolar, N. M., Burke, L. M., Hawley, J. A., Kassis, A., Karagounis, L. G., Li, K., King, C., Hellerstein, M., & Phillips, S. M. (2018). Effect of resistance training and protein intake pattern on myofibrillar protein synthesis and proteome kinetics in older men in energy restriction. *The Journal of Physiology, 596*, 2091–2120.
29. Wilkinson, D. J., Cegielski, J., Phillips, B. E., Boereboom, C., Lund, J. N., Atherton, P. J., & Smith, K. (2015). Internal comparison between deuterium oxide (D_2O) and L-[*ring*-$^{13}C_6$] phenylalanine for acute measurement of muscle protein synthesis in humans. *Physiological Reports, 3*, e12433.
30. Commerford, S. L., Carsten, A. L., & Cronkite, E. P. (1983). The distribution of tritium among the amino acids of proteins obtained from mice exposed to tritiated water. *Radiation Research, 94*, 151–155.
31. Price, J. C., Holmes, W. E., Li, K. W., Floreani, N. A., Neese, R. A., Turner, S. M., & Hellerstein, M. K. (2012). Measurement of human plasma proteome dynamics with (2)H(2)O and liquid chromatography tandem mass spectrometry. *Analytical Biochemistry, 420*, 73–83.
32. Burniston, J. G., Connolly, J., Kainulainen, H., Britton, S. L., & Koch, L. G. (2014). Label-free profiling of skeletal muscle using high-definition mass spectrometry. *Proteomics, 14*, 2339–2344.
33. Malik, Z. A., Cobley, J. N., Morton, J. P., Close, G. L., Edwards, B. J., Koch, L. G., Britton, S. L., & Burniston, J. G. (2013). Label-free LC-MS profiling of skeletal muscle reveals heart-type fatty acid binding protein as a candidate biomarker of aerobic capacity. *Proteomes, 1*, 290–308.
34. Hesketh, S., Srisawat, K., Sutherland, H., Jarvis, J., & Burniston, J. (2016). On the rate of synthesis of individual proteins within and between different striated muscles of the rat. *Proteomes, 4*, 12.
35. Xiao, G. G., Garg, M., Lim, S., Wong, D., Go, V. L., & Lee, W. N. (2008). Determination of protein synthesis in vivo using labeling from deuterated water and analysis of MALDI-TOF spectrum. *Journal of Applied Physiology (Bethesda, MD: 1985), 104*, 828–836.
36. Holmes, W. E., Angel, T. E., Li, K. W., & Hellerstein, M. K. (2015). Dynamic proteomics: In vivo proteome-wide measurement of protein kinetics using metabolic labeling. *Methods in Enzymology, 561*, 219–276.
37. Kasumov, T., Willard, B., Li, L., Sadygov, R. G., & Previs, S. (2015). Proteome dynamics with heavy water: Instruments, data analysis and biological applications. In S. Magdeldin (Ed.), *Recent Advances in Proteomics Research*. Rijeka: InTech.
38. Garlick, P. J., Millward, D. J., & James, W. P. (1973). The diurnal response of muscle and liver protein synthesis in vivo in meal-fed rats. *The Biochemical Journal, 136*, 935–945.
39. Lam, M. P., Wang, D., Lau, E., Liem, D. A., Kim, A. K., Ng, D. C., Liang, X., Bleakley, B. J., Liu, C., Tabaraki, J. D., Cadeiras, M., Wang, Y., Deng, M. C., & Ping, P. (2014). Protein kinetic signatures of the remodeling heart following isoproterenol stimulation. *The Journal of Clinical Investigation, 124*, 1734–1744.

40. Price, J. C., Guan, S., Burlingame, A., Prusiner, S. B., & Ghaemmaghami, S. (2010). Analysis of proteome dynamics in the mouse brain. *Proceedings of the National Academy of Sciences of the United States of America, 107*, 14508–14513.
41. Price, J. C., Khambatta, C. F., Li, K. W., Bruss, M. D., Shankaran, M., Dalidd, M., Floreani, N. A., Roberts, L. S., Turner, S. M., Holmes, W. E., & Hellerstein, M. K. (2012). The effect of long term calorie restriction on in vivo hepatic proteostatis: A novel combination of dynamic and quantitative proteomics. *Molecular and Cellular Proteomics, 11*, 1801–1814.
42. Kim, T. Y., Wang, D., Kim, A. K., Lau, E., Lin, A. J., Liem, D. A., Zhang, J., Zong, N. C., Lam, M. P., & Ping, P. (2012). Metabolic labeling reveals proteome dynamics of mouse mitochondria. *Molecular and Cellular Proteomics, 11*, 1586–1594.
43. Palmer, J. W., Tandler, B., & Hoppel, C. L. (1977). Biochemical properties of subsarcolemmal and interfibrillar mitochondria isolated from rat cardiac muscle. *The Journal of Biological Chemistry, 252*, 8731–8739.
44. Kavazis, A. N., Alvarez, S., Talbert, E., Lee, Y., & Powers, S. K. (2009). Exercise training induces a cardioprotective phenotype and alterations in cardiac subsarcolemmal and intermyofibrillar mitochondrial proteins. *American Journal of Physiology Heart and Circulatory Physiology, 297*, H144–H152.
45. Kasumov, T., Dabkowski, E. R., Shekar, K. C., Li, L., Ribeiro, R. F., Jr., Walsh, K., Previs, S. F., Sadygov, R. G., Willard, B., & Stanley, W. C. (2013). Assessment of cardiac proteome dynamics with heavy water: Slower protein synthesis rates in interfibrillar than subsarcolemmal mitochondria. *American Journal of Physiology Heart and Circulatory Physiology, 304*, H1201–H1214.
46. Shankaran, M., King, C. L., Angel, T. E., Holmes, W. E., Li, K. W., Colangelo, M., Price, J. C., Turner, S. M., Bell, C., Hamilton, K. L., Miller, B. F., & Hellerstein, M. K. (2016). Circulating protein synthesis rates reveal skeletal muscle proteome dynamics. *The Journal of Clinical Investigation, 126*, 288–302.
47. Holloway, K. V., O'Gorman, M., Woods, P., Morton, J. P., Evans, L., Cable, N. T., Goldspink, D. F., & Burniston, J. G. (2009). Proteomic investigation of changes in human vastus lateralis muscle in response to interval-exercise training. *Proteomics, 9*, 5155–5174.
48. Srisawat, K., Shepherd, S. O., Lisboa, P. J., & Burniston, J. G. (2017). A systematic review and meta-analysis of proteomics literature on the response of human skeletal muscle to obesity/type 2 diabetes mellitus (T2DM) versus exercise training. *Proteomes, 5*, 30.
49. Rennie, M. J., Smith, K., & Watt, P. W. (1994). Measurement of human tissue protein synthesis: An optimal approach. *The American Journal of Physiology, 266*, E298–E307.
50. Mitchell, C. J., Churchward-Venne, T. A., Parise, G., Bellamy, L., Baker, S. K., Smith, K., Atherton, P. J., & Phillips, S. M. (2014). Acute post-exercise myofibrillar protein synthesis is not correlated with resistance training-induced muscle hypertrophy in young men. *PLoS One, 9*, e89431.
51. Murphy, C. H., Churchward-Venne, T. A., Mitchell, C. J., Kolar, N. M., Kassis, A., Karagounis, L. G., Burke, L. M., Hawley, J. A., & Phillips, S. M. (2015). Hypoenergetic diet-induced reductions in myofibrillar protein synthesis are restored with resistance training and balanced daily protein ingestion in older men. *American Journal of Physiology Endocrinology and Metabolism, 308*, E734–E743.
52. Geiger, T., Velic, A., Macek, B., Lundberg, E., Kampf, C., Nagaraj, N., Uhlen, M., Cox, J., & Mann, M. (2013). Initial quantitative proteomic map of 28 mouse tissues using the SILAC mouse. *Molecular and Cellular Proteomics, 12*, 1709–1722.
53. Srisawat, K., Hesketh, K., Cocks, M., Strauss, J., Edwards, B. J., Lisboa, P. J., Shepherd, S., & Burniston, J. G. (2019). Reliability of protein abundance and synthesis measurements in human skeletal muscle. *Proteomics*, e1900194. https://www.ncbi.nlm.nih.gov/pubmed/31622029; https://onlinelibrary.wiley.com/doi/abs/10.1002/pmic.201900194
54. Lopez-Otin, C., Blasco, M. A., Partridge, L., Serrano, M., & Kroemer, G. (2013). The hallmarks of aging. *Cell, 153*, 1194–1217.

Part IV
Metabolomic

Chapter 10
Skeletal Muscle Metabolomics for Metabolic Phenotyping and Biomarker Discovery

Kenneth Allen Dyar, Anna Artati, Alexander Cecil, and Jerzy Adamski

10.1 Theory and History of Metabolomics

Metabolism is the process of chemical transformation within a biological context. Metabolites comprise all the small-molecule (<1 kDa) substrates and end products of metabolism, including sugars, nucleotides, lipids, amino acids, organic acids, ketones, aldehydes, amines, alkaloids, phenols, steroids, small peptides, xenobiotics, and drugs. Similar in scope to other high-throughput "omics" technologies, the aim of metabolomics is to comprehensively and unbiasedly detect, identify, and quantify the metabolome, i.e., the full complement of small molecules found in cells, biological fluids, or tissues. Here we present a brief introduction of how skeletal muscle metabolomics can be used for metabolic phenotyping and biomarker discovery.

K. A. Dyar (✉)
Division of Molecular Endocrinology, Institute for Diabetes and Cancer, Helmholtz Zentrum München, German Research Center for Environmental Health (GmbH), Neuherberg, Germany

German Center for Diabetes Research (DZD), Neuherberg, Germany
e-mail: kenneth.dyar@helmholtz-muenchen.de

A. Artati · A. Cecil
Research Unit Molecular Endocrinology and Metabolism, Helmholtz Zentrum München, German Research Center for Environmental Health (GmbH), Neuherberg, Germany

J. Adamski
Research Unit Molecular Endocrinology and Metabolism, Helmholtz Zentrum München, German Research Center for Environmental Health (GmbH), Neuherberg, Germany

Lehrstuhl für Experimentelle Genetik, Technische Universität München, Freising-Weihenstephan, Germany

German Center for Diabetes Research (DZD), Neuherberg, Germany

Department of Biochemistry, Yong Loo Lin School of Medicine, National University of Singapore, Singapore

J. G. Burniston, Y.-W. Chen (eds.), *Omics Approaches to Understanding Muscle Biology*, Methods in Physiology, https://doi.org/10.1007/978-1-4939-9802-9_10

The presence and abundance of a particular metabolite are determined by the sum of all genetic and environmental interactions [1]. At the cellular level, these include the presence and abundance of related metabolites and cofactors, protein expression and posttranslational modifications, gene expression, as well as epigenetic and genetic mechanisms. At higher levels of organization, these factors are further determined by specific physiological or pathophysiological contexts (time of day, age, sex, nutritional status, physical activity, stress, disease, medications), as well as environmental factors like diet composition, and host interaction with the gut microbiome. By providing a comprehensive metabolic snapshot of an organism within a particular context, metabolomics describes the phenotype of an organism better than any of the other high-throughput "omics" technologies [2].

Accordingly, metabolites are powerful diagnostic indicators (biomarkers), and they have been used to define health and disease since antiquity. For example, ancient Chinese and Hindu manuscripts describe doctors using the sweetness of urine (glycosuria) to diagnose diabetic patients already ~1500 BC [3]. Still today, measuring specific metabolites in urine or blood and comparing them to levels found in healthy individuals form the diagnostic basis of clinical chemistry, and metabolic profiling allows doctors to define health status, track disease progression, and chart the progress of various therapeutic interventions. This approach has successfully led to early detection and management of a host of metabolic diseases thanks to neonatal screening and regular health checkups.

The arrival of high-throughput metabolomics in the clinic quickly expanded this basic analytical framework. Due to its comprehensive nature, metabolomics captures more complex metabolic signatures that were previously unmeasured and thus unknown. This information has already allowed doctors to better stratify patient groups [4] and is expected to provide researchers with critical mechanistic insight into the etiology of multifactorial diseases [5]. An early indicator of the predictive potential of metabolomics is that some metabolic signatures can already predict future risk and therapeutic success for certain diseases, like type 2 diabetes, more than a decade in advance [1, 6]. It is anticipated that this kind of knowledge will quickly pave the way toward more personalized medicine.

In comparison, genome-wide association studies (GWASs) have similarly identified many genetic polymorphisms associated with disease risk [7–10]. However, GWASs struggle to provide the same predictive power as metabolomics. This is partly because effect size for polymorphisms is often small and so requires screening much larger populations to gain sufficient statistical power to identify disease-causing variants. This problem is only magnified for more complex polygenic and multifactorial diseases. On the other hand, different genetic variants, whether within the same locus or in distinct and seemingly unrelated loci, can result in the same or related metabolic phenotypes. Screening many intermediate phenotypes with metabolomics therefore gives a broader picture of how different metabolic pathways may be related and even relevant for the same disease [11]. To harness the best of both worlds, metabolomics can also be combined with GWASs and other high-throughput "omics" strategies. This combined approach can provide novel biological insight and biomarkers for a wide range of heritable diseases and has already uncovered new gene-metabolite associations relevant for cardiovascular disease and mood disorders [6].

The equipment needed to perform metabolomics studies originated in the 1950s by synergizing complementary analytical and computational techniques [12–16]. Coupling the physical separation capabilities of different chromatography methods with the precision and accuracy of mass spectrometers made it possible to confidently identify and quantify metabolites with high sensitivity and high specificity over a wide range of applications.

Modern clinical metabolomics was born for the first time in 1971, with Horning and Horning [17]; Mamer, Crawhall, and Tjoa [18]; Pauling, Robinson, and colleagues [19] profiling hundreds of urine metabolites. Around the same time, advances in nuclear magnetic resonance (NMR) spectroscopy allowed for detection and quantification of ATP, phosphocreatine, and inorganic and sugar phosphates in intact skeletal muscles [20]. NMR was quickly adopted to further investigate metabolites in different muscle types under aerobic or anaerobic conditions, during electrical stimulation, and in diseased patients with Duchenne muscular dystrophy or *spinal muscular atrophy* [21]. The field of metabolic profiling continued to grow over the following decades, alongside continued technological advances in NMR, mass spectrometry, chromatography, and computing. However, the explosive "rebirth" of metabolomics as we know it today evolved from the rapid advances in biostatistics, bioinformatics, and systems biology during the early genomics era.

The most common metabolomics experiments involve non-targeted (discovery) or targeted (validation) approaches. Non-targeted metabolomics provides a global view of all detectable metabolites within a biological sample. Extraction and analytical methods for non-targeted metabolomics are generally designed and optimized to detect the widest range of biochemical classes with the largest number of individual metabolites possible. This allows for the relative quantification of "known" metabolites, i.e., those already identified and registered in metabolite databases, but also "unknown" metabolites, which are detected yet still unidentified. These unidentified metabolites may be important for phenotypic investigations and can be further validated and identified in follow-up studies using a set of characteristic parameters (i.e., retention times and mass spectra). Targeted metabolomics experiments allow for the absolute quantification of a preselected set of metabolites, and so by nature, these studies are more focused, with extraction methods optimized around specific classes and panels of metabolites. However, in special cases non-targeted methods are similarly adapted to focus on more specific metabolite classes, i.e., lipidomics for lipids or glycomics for sugar metabolites.

Successful metabolomics studies rely on proper experimental design, standardized sample processing, suitability of analytical methods, and biostatistics-bioinformatics data processing and analysis (see Fig. 10.1 for an overview of a typical metabolomics workflow).

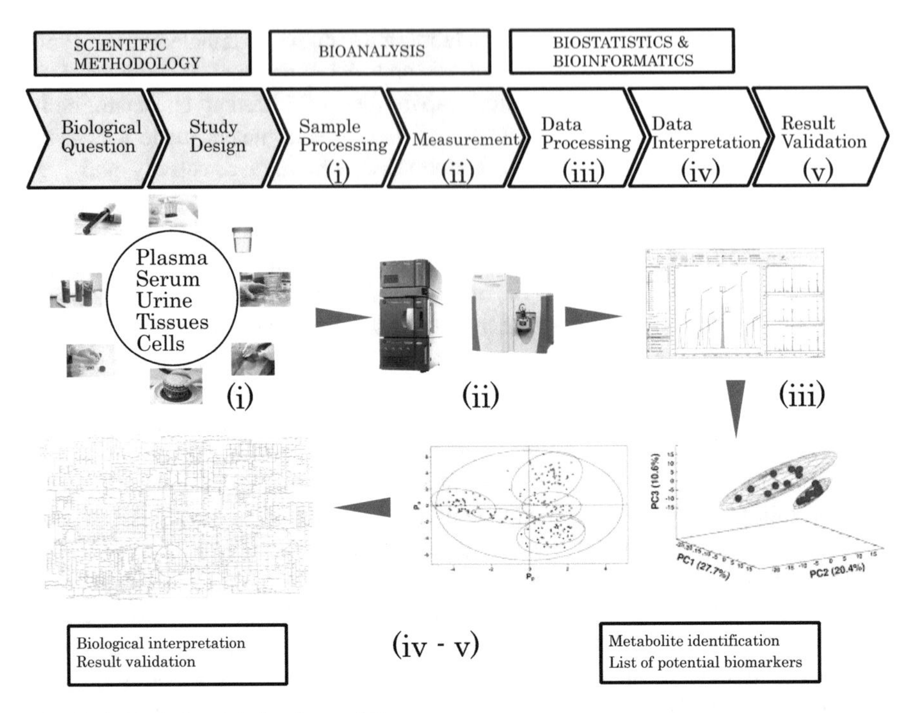

Fig. 10.1 Typical metabolomics workflow

10.1.1 Experimental Design

Study design for metabolomics experiments normally follows two distinct strategies [22]. The first strategy involves a well-controlled experimental environment, with the only variable being the experimental treatment or condition. This usually results in large changes in the metabolome, so these types of experiments do not normally require a large number of replicates to obtain sufficient statistical confidence. Examples are in vitro experiments, cultured tissue systems, and animal studies. The second strategy involves population or epidemiological studies in which metabolite changes are often more subtle. In this case, many more samples may be required to provide statistical confidence and gain an understanding of metabolic status.

Quality control (QC) and quality assurance (QA) are critical in metabolomics experiments, especially in large-scale studies where it is often impossible to run all samples in a single analytical batch. Instrument reproducibility and suitability must also be considered and carefully followed throughout each study. Furthermore, QC samples are required in every batch in order to provide robust QA for each detected metabolic feature, as QC responses are used as the algorithmic basis for assessing data quality in later data processing steps.

10.1.2 Sample Processing

The first crucial step in sample processing is sample collection. Some metabolites are extremely labile, so cellular metabolism must be immediately quenched. For tissue samples, this can be achieved by ultrarapid freeze-clamping in situ [23] or by snap freezing samples immediately after collection in liquid nitrogen. In either case, samples are subsequently stored at −80 °C until further processing. It has been shown that long-term storage of some samples, like plasma, can alter the concentrations of some metabolites and thus bias data analysis [24]. This is a major concern for longitudinal studies, in which samples may be stored, processed, and measured over several years or even decades. There may be differences in the long-term stability of metabolites depending on the type of sample, whether plasma, serum, urine, or different tissues; however, the literature regarding this point is extremely sparse or nonexistent.

Sample preparation prior to analytical measurement typically involves metabolite extraction, protein precipitation, and optional chemical derivatization. These processes must all be optimized and validated for each metabolite of interest.

10.1.3 Analytical Method

As mentioned previously, modern metabolomics has been fueled by major technological improvements in instrumentation. The most frequently used technologies in metabolomics are mass spectrometry (MS) and NMR. Resolution with either type of analysis is greatly enhanced when separation steps are introduced prior to metabolite detection. Common separation steps include gas chromatography (GC), capillary electrophoresis (CE), liquid chromatography (LC), or ultra-high-performance liquid chromatography (UHPLC), which can be coupled to either MS or NMR [25].

Analytical methods for targeted metabolomics must likewise first be validated for all selected biochemicals in a study. The most commonly applied techniques are based on GC-MS, LC-MS, UHPLC-MS, or flow injection assay mass spectrometry (FIA-MS) [26, 27]. Since non-targeted metabolomics analytical methods must be suitable for a wide range of metabolite classes, these studies often require a high degree of parallel analytical platforms (i.e., multiple analyses on LC- and GC-MS). This makes it easier to cover as many metabolites as possible and to avoid biases for specific chemical classes. The most commonly used platforms are NMR, CE-MS, GC-MS, LC-FT-ICR-MS, or UHPLC-MS [28, 29]. Special algorithms for metabolite identification with specific databases are also required [30–32].

10.1.4 Data Preprocessing

The raw data of mass spectrometry and NMR-based metabolomics experiments are mass spectra peaks. These peaks must first be assigned a metabolite identity based on reference spectra contained within large chemical databases. There are several different computational approaches and tools for identifying and annotating metabolites, and the reader is encouraged to consult [15] for more detailed information. Once metabolite identities are assigned, data preprocessing and analysis can begin.

In targeted metabolomics, data analysis begins by calculating whether the measured metabolite concentrations lie within the corresponding lower limit of quantification (LLOQ) and upper limit of quantification (ULOQ). These end points of the linear range of metabolite concentrations are determined during assay development and are specific for the MS machines used to measure the samples. The most commonly used quality control (QC) samples are human plasma reference samples or National Institute of Standards and Technology (NIST) samples. These are used to determine the coefficient of variation (CV) during the measurement process. Metabolites showing more than 25% CV in their QC samples are discarded, since there is a high likelihood they were incorrectly measured.

These first validation steps are very stringent. However, the range of acceptable metabolite measurements can be broadened to also accept metabolite measurements below the LLOQ—down to the absolute limit of detection (LOD), below which the chosen mass spectrometer is unable to reliably detect peaks. Reference samples are not only used for quality control but also to normalize for inter-batch effects when measuring a high number of samples split into different batches.

In a typical targeted metabolomics experiment, plate-specific mean values ("plate means") are calculated for each plate and each metabolite:

$$\text{Platemean}[\text{metabolite}(x)] = \frac{\sum C[\text{metabolite}(x)]}{N}$$

$$\text{where:}$$

$$N = \text{number of reference samples}$$

$$C = \text{metabolite concentration}$$

The plate means are then used to calculate the overall "mean of all plates":

$$\text{Overall mean}[X] = \frac{\sum \text{Means}[X]}{N}$$

$$\text{where:}$$

$$X = \text{metabolite}(x) \text{ of Plates}(n \ldots n_j)$$

$$N = \text{number of Plates}(n \ldots n_j)$$

From the overall mean, the plate factors are calculated:

$$\text{Factor}[Y] = \frac{\text{Overall mean}[X]}{\text{Mean}[Y]}$$
$$\text{where:}$$
$$X = \text{metabolite}(x) \text{ of Plates}(n \ldots n_j)$$
$$Y = \text{Plate}(n)[\text{metabolite}(x)]$$

In the final normalization step, metabolite concentrations are multiplied by the corresponding plate factor, e.g., metabolite concentrations of each sample on plate 1 are multiplied with plate factor 1:

$$\text{Normalized}[Y] = \text{Factor}[Y] * \text{Concentration}[Y]$$
$$\text{where:}$$
$$Y = \text{Plate}(n)[\text{metabolite}(x)]$$

Non-targeted metabolomics measurements cannot be validated in the same way as targeted assays. Since reference samples are normally not included, parameters like LLOQ, ULOQ, and LOD are not available. However, it is possible to generate pooled QC samples from all the samples to be measured. These samples can then be used in much the same way as the reference samples from targeted metabolomics.

For both approaches, and especially for non-targeted metabolomics, “NA” values are a common occurrence. A result of “NA” in a metabolomics measurement does not mean “no metabolite present” but rather “no metabolite concentration detectable with the given method.” There are several ways to deal with these occurrences using data imputation [33–38]. One of the most commonly used data imputation methods is to replace all “NA” values by half the minimum value measured for a given metabolite. Other approaches use the LOD instead of the minimum value and divide the LOD either by 2 or by $\sqrt{2}$. Some other imputation methods use zero value or median value imputation, k-nearest neighbors, or imputation by machine learning algorithms such as random forest. Whichever way imputation is handled, it is important to be consistent and transparent and explicitly state the impact of imputation (how many metabolites were imputed, what kind, etc.) in the methods. If many “NA” values are present, it is prudent to define exclusion criteria. Commonly used ranges for metabolite exclusion are between 30% and 50% of “NA” values. After initial data preprocessing, the data can be checked for outliers. This is commonly done using box plots, with every metabolite above or below the 1.5 inter quartile range considered an outlier, and thus discarded. Again, exclusion criteria should be specified in the methods.

10.1.5 Data Analysis

After preprocessing, the data can be evaluated with many statistical methods and tools [39–54]. The first step is to determine which of the measured metabolites are of interest: one specific metabolite, groups of metabolites, or all? Depending on the experimental setup and data distribution, either uni- or multivariate analyses can be conducted. Data distribution can be determined using Shapiro-Wilk tests, QQ plotting, or bootstrapping, with results determining which downstream statistical tests are appropriate.

Univariate tests include *t*-tests, Mann-Whitney *U* tests, Kruskal-Wallis tests, as well as ANOVA. For the multivariate tests, a popular option is unsupervised principal component analysis (PCA). This method does not take group information into account and is based on the assumption that variables (in this case metabolites) with a high variance carry a high amount of information, and conversely variables with low variance convey less information. PCA allows for a quick overview of the data composition, as samples with similar metabolite concentrations tend to cluster together in the PCA plot.

In contrast, a partial least squares discriminant analysis (PLS-DA) utilizes a supervised approach taking group information into account [39, 41, 43, 53, 55–58]. The principal components of a data matrix (metabolite measurements) and their respective response vector (group information, also called "discriminant") are calculated in order to maximize their covariance. Prior knowledge of which groups are in the data set is therefore paramount for PLS-DA. Orthogonal PLS-DA (OPLS-DA) takes this one step further by removing orthogonal noise unrelated to the response vector. An OPLS-DA is therefore better able to differentiate groups than a PLS-DA. There are some important details to keep in mind when performing PLS-DA or OPLS-DA. For one, data overfitting is detected by the R^2 and Q^2 values. The R^2 value is a measure for the goodness of fit of the measured data related to the calculated regression of the data, and the Q^2 value is a measure for the difference of the sample groups. These should always be positive. It is also important to perform permutation tests to check whether sample grouping is specific for the tested phenotypes, i.e., whether the same clustering of the data is observed after grouping information is randomly permuted n-times (most commonly $n = 2000$). If the clustering is observable in more than 5% of the cases, the test is considered to have failed and the PLS-DA to be invalid.

Both PCA and PLS-DA yield a list of variables responsible for clustering data in the PCA- or PLS-DA plots. In the PCA, the loading for each principal component can be extracted, whereas for PLS-DA, "variables of importance" fulfill this role. Variables of importance above a value of 1.0 are considered significant and used to narrow down the list of relevant metabolites. From the PCA, one can extract the relevant loadings by checking whether individual contributions are in the sloped area of the curve or whether they are asymptotic, and contributions of the related metabolites are therefore uniform.

Using data from the statistical tests and/or the results from PCA/PLS-DA, the search for possible biomarkers can begin. This is most often achieved by constructing a receiver operating characteristic (ROC) curve [59], which is a graphical representation of a so-called binary classifier system (see Fig. 10.2). When comparing biological systems for biomarker detection, ROC curves are often superior to using fold changes or odds ratios, as the ROC provides information on both the sensitivity and selectivity of each biomarker. For this end, the data is split into

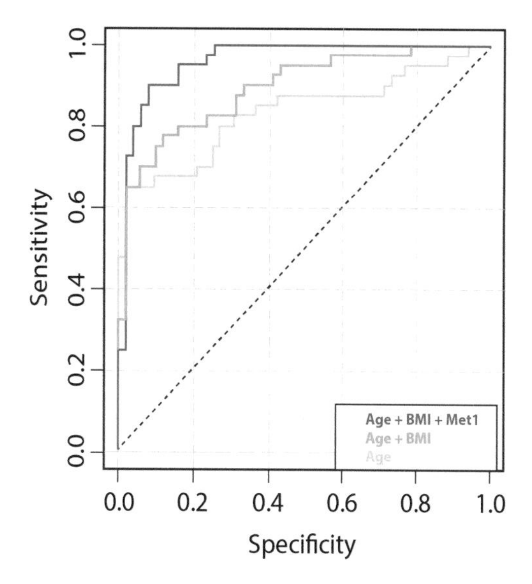

Fig. 10.2 Example of a receiver operating characteristic (ROC) curve. A ROC curve reveals specificity and sensitivity of models used to diagnose or differentiate data sets. In this example three different models were built based only on age of subjects (yellow), age and body mass index (BMI, green), or age, BMI, and a biomarker metabolite (Met1, blue). The model with age, BMI, and metabolite shows the best performance with 84% specificity and 90% sensitivity. The dashed line depicts performance of a random approach with 50% specificity and 50% selectivity. Figure redrawn using data from [60] with permission from Oxford University Press

two groups, based on binary classification, e.g., treated and untreated samples, or healthy versus diseased. For this binary classification, the true positive, also called "sensitivity," and true negative rates, also called "specificity" (TPR and TNR, respectively), can be calculated as follows:

TPR

$$\frac{\text{True positive}}{\text{True positive} + \text{False negative}}$$

TNR

$$\frac{\text{Falsepositive}}{\text{False positive} + \text{True negative}}$$

The ROC curve is obtained by plotting these two values against each other for pre-defined thresholds (T) on the axes:

$$\text{TPR}(T) = \int_T^{inf} f_1(x)dx \text{ versus } \text{FPR}(T) = \int_T^{inf} f_0(x)dx$$
$$\text{where FPR} = 1 - \text{TNR}$$

Every point on the ROC curve represents a specificity/sensitivity pair for the chosen criterion. From here, the area under the curve (AUC) can be calculated as a measure of performance of the ROC. It can also be used to compare different ROCs.

There is also a large array of additional tools that can be implemented for downstream metabolomics data analysis, interpretation, visualization, and integration with other "omics" data [61]. These rely on data associations and current knowledge about biochemical pathway structure and regulation for pattern recognition, network inference and modeling, pathway activity prediction, metabolic pathway enrichment, and overrepresentation. The reader is encouraged to consult [15] for a recent overview of these and other useful approaches.

10.2 Major Applications

Skeletal muscle plays a highly dynamic role in metabolism and whole-body energy homeostasis, both at rest [62] and during exercise [63]. Accordingly, skeletal muscle's role in health and disease has been an active research target since the very first metabolomics experiments. This has inevitably produced a rich tapestry of muscle metabolomics studies, briefly summarized in [64].

However, the vast majority of these studies have targeted muscle indirectly by measuring metabolites in blood, urine, or even interstitial fluid [65, 66] after various exercise protocols, under different nutritional challenges, or under different diseased states. These studies include elegant and informative arterial-venous blood sampling across the vascular bed of human forearm muscles [67], allowing for quantitative characterization of dynamic uptake and release of muscle metabolites. Further applications of these indirect approaches under different experimental conditions will surely continue to uncover novel signaling molecules and regulatory mechanisms associated with muscle function and disease pathogenesis.

On the other hand, muscle metabolomics experiments focusing directly or exclusively on muscle tissue, rather than associated body fluids, are not as prevalent in the literature. These can be broadly divided among five main research objectives:

1. Technical studies centered around various extraction and analytical techniques [64] and their impact on muscle metabolite recovery and detection.
2. Studies aimed at understanding basic muscle physiology, including exercise and nutritional studies focused on muscle development and growth, adaptation, performance, and metabolism [68–74].
3. Studies investigating muscle metabolites during disease, disuse, or injury and the effects of various interventions [75–77].
4. Studies performed for diagnostics, biomarker discovery, and forensics, including time of death estimation [78–86].
5. Metabolic phenotyping of animal models and cell culture systems for basic research [87–96].

A comprehensive overview of all muscle metabolomics studies is outside of the scope of this brief chapter, and this short survey of muscle metabolomics experiments is by no means complete. It is merely meant to give a brief overview of the wide range of applied muscle metabolomics studies already in the literature, and we apologize to any colleagues whose work we have neglected due to space constraints. In the following section, we will highlight only a few examples of how time series metabolomics can help circumvent a limitation of many metabolomics experiments, and can give important insight into disease mechanisms.

Complex metabolic diseases can be characterized according to their unique metabolic signatures, i.e. how specific panels of metabolites change relative to healthy individuals. As metabolism is a dynamic process, dynamic metabolic signatures may likewise appear only transiently within precise temporal and spatial windows that are difficult to know a priori. Dynamic metabolic signatures may thus escape detection when sampling a single tissue at a single time point. In a seminal paper, Koves and colleagues used targeted metabolomics to characterize metabolic signatures associated with insulin resistance in muscles from rats and mice [97]. By comparing both fed and fasted lean rats fed a standard chow diet with fed and fasted obese rats fed a high-fat diet, the authors revealed a transient metabolic signature of persistently increased fatty acylcarnitines in muscles and serum from obese and diabetic rats during the fed state. This was accompanied by increased rates of incomplete β-oxidation and reduced levels of organic acid intermediates associated with glycolysis and the TCA cycle, including lactate, citrate, malate, and succinate. Their data revealed impaired metabolic flexibility in the fasted-to-fed transition and suggested a causal link between mitochondrial lipid overload and muscle insulin resistance and glucose intolerance. More recently, time-of-day dependent metabolite changes in human vastus lateralis muscle were revealed in response to high-fat and high-carbohydrate diets [72] underscoring the importance of diet composition and time-of-day in determining the muscle metabolome of humans.

Temporal perspectives of metabolite dynamics can be further expanded to better capture more complex and dynamic metabolic signatures and relationships under a typical range of physiological conditions, including feeding-fasting and rest-activity cycles (see Fig. 10.3). This has been achieved by sampling tissues, cells, or organelles at high temporal resolution and by comparing metabolite dynamics across 24 h under different nutritional, pharmacological, genetic, or exercise interventions [99–107].

This 24-h metabolomics perspective was recently applied to serial biopsies from human vastus lateralis muscles, and in primary human myotubes after circadian clock disruption using targeted lipidomics [108], and via non-targeted metabolomics in murine tibialis anterior muscles after muscle-specific clock disruption [109, 110]. These studies have provided important mechanistic insight into the multiple roles muscle clocks play in maintaining glucose, lipid, and amino acid homeostasis over 24 h. They have also shown how muscle clock disruption is associated with local and systemic metabolic disturbances, including well-known risk factors for metabolic diseases like intramuscular accumulation of bioactive lipids and muscle insulin resistance.

muscle diurnal substrate selection

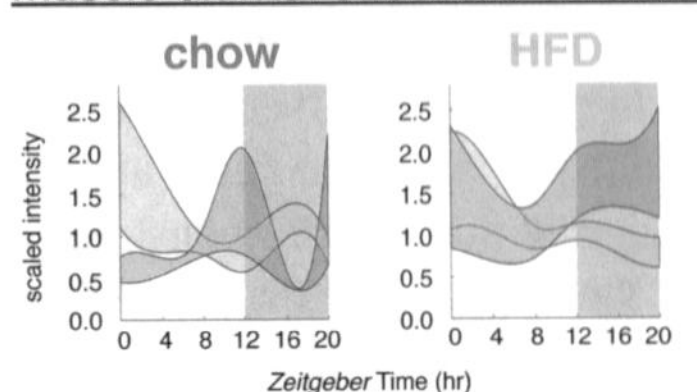

plotted muscle metabolites

glycolytic intermediates
glucose
glucose 6-phosphate
fructose 1,6-diphosphate
dihydroxyacetone phosphate
3-phosphoglycerate
phosphoenolpyruvate
pyruvate

long & medium-chain acylcarnitines
linoleoylcarnitine
oleoylcarnitine
stearoylcarnitine
palmitoylcarnitine
myristoleoylcarnitine
myristoylcarnitine
laurylcarnitine
decanoylcarnitine
octanoylcarnitine
hexanoylcarnitine

Fig. 10.3 Skeletal muscle diurnal fuel selection is altered by diet composition. Non-targeted 24-h metabolomics profiling using LC/MS performed on gastrocnemius muscles collected from mice under controlled laboratory conditions after 10 weeks of ad libitum feeding a standard chow diet or a high-fat diet (HFD). The 12-h light-dark cycle of the animal facility is indicated by white or gray shaded areas, respectively, with "lights on" at *zeitgeber* time 0 and "lights off" at *zeitgeber* time 12. Green and pink plots show the range (colored area) between minimum and maximum abundance for the muscle metabolites indicated on the right. Cubic spline interpolation was used to estimate continuous abundance. Reprinted from [98], with permission from Elsevier

This is because muscle metabolism plays an integral role in metabolic homeostasis in both the fed and fasted states. For example, muscle is the main site for insulin-stimulated glucose uptake [111] and a major consumer of lipids [112]. It also plays an important role in whole-body protein metabolism [113] as a major destination for amino acids after a meal [114] and as the main source of circulating amino acids during periods of starvation and insulin deficiency [115]. What happens in muscle thus impacts other tissues, including the brain, via metabolic cross-talk [116] and by the production and secretion of myokines [117].

A comprehensive metabolomics atlas of circadian metabolism recently underscored this interconnected role of muscle circadian metabolism and organ cross-talk. Profiling eight different mouse tissues, including gastrocnemius muscle, across 24 h under standard chow diet or high-fat diet [98], 24-h metabolite temporal associations were reconstructed both within and across different tissues. This dynamic temporal and spatial overview of the metabolome revealed novel metabolic cross-talk between tissues and suggested novel pathogenic relationships. For example, high-fat diet was associated with increased muscle production and release of 3-methylhistidine and alanine, both established biomarkers for muscle protein degradation. These were temporally associated with increased glycerol 3-phosphate levels in liver, suggesting a link between muscle protein turnover under high-fat diet and increased liver production and secretion of glyceride-glycerol. While tracer studies are needed to independently validate this particular relationship, muscle and serum alanine were also found transiently elevated when profiling across 24 h in insulin-resistant mice with muscle-specific clock disruption [110] and in serum of insulin-resistant humans [118]. Together, these data suggest alanine as a novel risk biomarker for type 2 diabetes (see Fig. 10.4). However, since this signal appeared transiently over several hours, it may have been missed if sampling muscle or blood at the "wrong" time or under less controlled conditions.

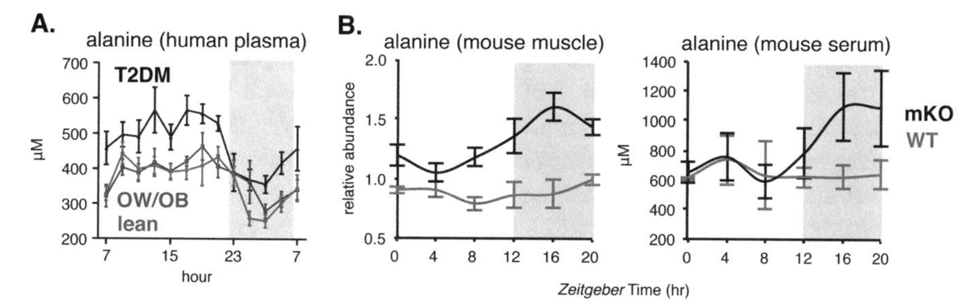

Fig. 10.4 Biomarker discovery using 24-h metabolomics profiling. (**a**) Targeted 24-h metabolomics profiling of human plasma under controlled laboratory conditions using LC/MS revealed elevated alanine among several novel biomarkers distinguishing overweight/obese type 2 diabetes mellitus subjects (T2DM) from age- and BMI-matched overweight/obese (OW/OB, blue) and age-matched lean subjects (red). Lean and OW/OB groups were also matched for glucose homeostasis markers (HbA1c, fasting glucose and insulin). Alanine was elevated in T2DM subjects during the last half of the sleeping/fasting phase (gray region) and throughout the awake/feeding phase (white region). Reprinted from [118] under the terms of the Creative Commons Attribution 4.0 International (CC BY 4.0). (**b**) Non-targeted 24-h metabolomics profiling of mouse muscles under controlled laboratory conditions using GC/MS similarly revealed elevated alanine in tibialis anterior muscles from muscle-specific *Bmal1* knockout mice (mKO) compared to wild-type (WT) littermates. Alanine was elevated in muscles from insulin resistant mKO mice at the end of the sleeping/fasting phase (white region) and throughout the awake/feeding phase (gray region). Targeted 24-h metabolomics profiling of mouse serum using GC/MS likewise detected elevated serum alanine in mKO mice during the awake/feeding phase. Adapted from [110] under the terms of CC BY 4.0

10.3 Limitations and Extensions of Metabolomics

Metabolomics is a tool, and like any analytical tool, it was designed for specific applications and has certain technical limitations worthy of consideration. For one, the small molecules detected by metabolomics span from 20 to 1000 Da and include a wide range of chemical structures and properties. Accordingly, there is no single optimum extraction protocol and/or analytical method to detect all metabolites within a sample. Combining different extraction techniques and different analytical methods can help to circumvent this issue, but this is not always an option.

Metabolite concentrations also span a wide dynamic range, from picomolar to millimolar, sometimes complicating their detection and discrimination. Another consideration is that some metabolites are chemically and thermally labile, making their recovery extremely challenging. One particular issue for non-targeted metabolomics experiments is determining the absolute number of unknown metabolites, and one should keep in mind that lack of detection does not necessarily equate with absence. Instrument variation can also be problematic, especially when measuring a large number of samples in different batches or over different days, and assay stability must be closely monitored throughout each experiment to ensure data quality.

Minimizing confounding factors is essential for producing meaningful metabolomics data and for interpreting results. Failure to properly control for these will negatively impact the outcome of the study and can result in misdiagnosis or identification of fewer biomarkers. This is particularly relevant for human metabolomics studies, which inherently contain more genetic and environmental variations than cell culture and animal studies. Sources of sample heterogeneity in human studies can result from individual variation; differences in sampling time, place, or condition; population differences; and factors like age, sex, health status, and environmental exposures. It may be difficult to control for every confounding factor; nevertheless they should be accounted for during data analysis.

Even after controlling for confounding factors, analytical and biological variation can still occur at multiple steps during sample processing, including during collection, storage, or extraction. Haid and colleagues [24] reported significant changes in stability of a wide range of metabolite classes when human plasma samples were stored long term at −80 °C. They found that storage of plasma samples at −80 °C for 5 years leads to altered concentrations of some amino acids, acylcarnitines, glycerophospholipids, sphingomyelins, and the sum of hexoses.

Many biomarker studies may also be limited in the amount of mechanistic insight they can provide regarding disease pathogenesis and treatment. This is because most predictive risk metabolites have been measured in blood or urine from patients, generally under specific controlled conditions (i.e., in the morning after an overnight fast). Many biomarkers may thus not adequately reflect the primary diseased tissue nor the particular circumstances under which pathogenic mechanisms arise (e.g., prolonged hyperglycemia after a meal). Metabolic signatures in blood and urine can differentiate diseases and reflect crucial aspects of tissue function and pathology. However, they are not a substitute for directly measuring the diseased tissue, which may provide more relevant mechanistic information. Accordingly, many current predictive risk biomarkers may only reflect indirect consequences of the diseases they describe rather than direct causes, thus limiting their utility in disease prevention or maintenance.

Finally, metabolism is a dynamic process, with metabolite concentrations determined by a balance between their production and consumption. As mentioned previously, metabolomics experiments performed at a single time point can thus only give a static snapshot of this balance. Time series metabolomics experiments can provide additional descriptive information about how metabolites and metabolic pathways are related and change over time, but cannot provide quantitative data regarding pathway activity and rates of metabolic flux. Metabolite concentrations and metabolic pathway flux are not always in agreement [119], and the only way to accurately infer pathway activity is to complement metabolomics studies with flux analysis using isotope tracers. For additional details on flux analysis, the reader is encouraged to consult some useful recent reviews [15, 119].

10.4 Vital Future Directions

10.4.1 Metabolomics for Precision/Personalized Medicine

Precision medicine involves considering a patient's genotype and phenotype when designing an optimal treatment strategy [120]. Genetic background and gene expression do not always track with health status, yet metabolomics provides a quantifiable readout of a patient's current biochemical state. Metabolomics will thus continue to be a growing and essential part of precision medicine's toolbox. Today most clinicians collect only a small fraction of available metabolite information, such as glucose concentration to monitor diabetes, cholesterol and high-density lipoprotein/low-density lipoprotein ratio to assess cardiovascular health, or blood urea nitrogen (BUN) and creatinine for kidney function. The narrow range of diagnostic analytes currently in use will be rapidly complemented by more comprehensive metabolomics analyses. As metabolomics enters into the field of clinical testing and more data becomes available, a wider range of biochemical signatures reflecting a wider spectrum of health, disease, and therapeutic potential will become available. The full emergence of clinical metabolomics is currently limited by a number of specific challenges, including method standardization, data analysis and interpretation, quality control, and the need for rigorous analytical validation [121]. However, as these and other challenges continue to be addressed, clinical metabolomics is expected to play an ever-increasing role in clinical diagnostics and decision-making [122].

10.4.2 Predictive and Preventive Health: Supporting the Worldwide Healthcare System [123]

Chronic diseases, including diabetes, cardiovascular disease, respiratory diseases, and cancer, are currently among the leading causes of death. They also create a major financial burden on economies throughout the world. In combination with promoting a healthy lifestyle (proper nutrition and exercise), monitoring and maintaining normal values for key health metrics like blood pressure, blood glucose, and cholesterol and lipid status greatly reduces disease risk. Integrating more comprehensive data from genomics, metabolomics, and other "omics" approaches will provide more comprehensive and personalized health metrics, along with novel predictive risk biomarkers. These will improve our potential for prevention in the earliest stages of disease and allow for better prediction and tracking of health status, thus improving overall quality of life and significantly reducing the financial burden from chronic diseases. These benefits are urgently needed, as the global population continues to rapidly become older [124] and more obese [125], thus increasing the general risk for chronic diseases.

10.4.3 Single Muscle Fiber Metabolomics

Metabolomics has been applied and validated in a variety of biological matrices, including cultured cells, body fluids, and tissues. However, even cultured cells with identical genotypes display heterogeneous phenotypes due to differences in cellular dynamics and unique microenvironments. Likewise, skeletal muscle is a highly heterogeneous tissue, with each muscle's functional and metabolic characteristics determined by its fiber type composition. Muscle fiber type is determined by intrinsic fiber-specific properties and by extrinsic factors, including hormones and motor neuron activity patterns [126]. Different muscle fiber types can be distinguished across a metabolic spectrum according to their predominant ATP source, with oxidative fibers relying more on oxidative phosphorylation and glycolytic fibers relying more on anaerobic glycolysis. However, in addition to quantitative differences in mitochondrial content, cutting-edge single muscle fiber proteomics recently revealed fiber-specific qualitative differences among mitochondrial proteomes [127], reflecting mitochondrial specialization among different fiber types [128].

Single muscle fiber metabolomics is similarly poised to unravel a previously hidden level of cellular heterogeneity and complexity among different fiber types. This holds obvious promise for providing a deeper understanding of muscle metabolism in health and disease. Single cell metabolomics is currently conducted using a shotgun-like MS approach with high mass resolution [129]. However, single-cell metabolomics is still in its infancy, and many technical challenges remain that are not so uncommon for the other single cell "omics" approaches. These include low sample quantity, and rapid turnover of cellular metabolites during fiber isolation, which may ultimately limit detection to only the most abundant and most resilient metabolites. As analytical techniques and sensitivity continue to improve, comprehensive, high-resolution single muscle fiber metabolomics seems just on the horizon.

10.4.4 Spatial Resolution in Metabolomics Experiments

A current limitation for most metabolomics experiments is that metabolites are normally extracted from whole homogenates of tissues and cells, losing all information about cell type-specific or subcellular metabolite localization. One way researchers have circumvented this issue is by performing fractionation of different subcellular compartments prior to metabolomics measurements [101, 130], but additional processing steps may delay quenching and lead to some metabolic drift. Furthermore, depending on tissue composition, one may likewise start with a mixture of several different cell types, again losing important information about cell type-specific metabolic heterogeneity.

Mass spectrometry imaging (MSI) is one promising technique to avoid these issues, as it directly visualizes the spatial distribution of metabolites in tissue sections or cells. While this technique is generally considered descriptive, the signal generated is proportional to analyte abundance. A variety of ionization methods are currently employed for MSI, such as matrix-assisted laser desorption/ionization (MALDI), desorption electrospray ionization (DESI), and laser ablation electrospray ionization (LAESI). For example, MALDI-MSI has been used to visualize and distinguish metabolites in different muscles and fiber types of rats and mice after different diet and exercise regimens [69, 131–134]. Continued technological advances that further increase sensitivity and spatial resolution should soon make it possible to confidently detect and quantify hundreds of subcellular metabolites in situ, opening up a whole new level of biochemical complexity and mechanistic understanding.

References

1. Holmes, E., Wilson, I. D., & Nicholson, J. K. (2008). Metabolic phenotyping in health and disease. *Cell, 134*, 714–717.
2. Patti, G. J., Yanes, O., & Siuzdak, G. (2012). Innovation: Metabolomics: The apogee of the omics trilogy. *Nature Reviews Molecular Cell Biology, 13*, 263–269.
3. van der Greef, J., van Wietmarschen, H., van Ommen, B., & Verheij, E. (2013). Looking back into the future: 30 years of metabolomics at TNO. *Mass Spectrometry Reviews, 32*, 399–415.
4. Nicholson, J. K. (2006). Global systems biology, personalized medicine and molecular epidemiology. *Molecular Systems Biology, 2*, 52.
5. Ramautar, R., Berger, R., van der Greef, J., & Hankemeier, T. (2013). Human metabolomics: Strategies to understand biology. *Current Opinion in Chemical Biology, 17*, 841–846.
6. Newgard, C. B. (2017). Metabolomics and metabolic diseases: Where do we stand? *Cell Metabolism, 25*, 43–56.
7. McCarthy, M. I., Abecasis, G. R., Cardon, L. R., Goldstein, D. B., Little, J., Ioannidis, J. P., & Hirschhorn, J. N. (2008). Genome-wide association studies for complex traits: Consensus, uncertainty and challenges. *Nature Reviews Genetics, 9*, 356–369.
8. Samani, N. J., Erdmann, J., Hall, A. S., Hengstenberg, C., Mangino, M., Mayer, B., Dixon, R. J., Meitinger, T., Braund, P., Wichmann, H. E., Barrett, J. H., Konig, I. R., Stevens, S. E., Szymczak, S., Tregouet, D. A., Iles, M. M., Pahlke, F., Pollard, H., Lieb, W., Cambien, F., Fischer, M., Ouwehand, W., Blankenberg, S., Balmforth, A. J., Baessler, A., Ball, S. G., Strom, T. M., Braenne, I., Gieger, C., Deloukas, P., Tobin, M. D., Ziegler, A., Thompson, J. R., Schunkert, H., & Wtccc, and the Cardiogenics Consortium. (2007). Genomewide association analysis of coronary artery disease. *The New England Journal of Medicine, 357*, 443–453.
9. Todd, J. A., Walker, N. M., Cooper, J. D., Smyth, D. J., Downes, K., Plagnol, V., Bailey, R., Nejentsev, S., Field, S. F., Payne, F., Lowe, C. E., Szeszko, J. S., Hafler, J. P., Zeitels, L., Yang, J. H., Vella, A., Nutland, S., Stevens, H. E., Schuilenburg, H., Coleman, G., Maisuria, M., Meadows, W., Smink, L. J., Healy, B., Burren, O. S., Lam, A. A., Ovington, N. R., Allen, J., Adlem, E., Leung, H. T., Wallace, C., Howson, J. M., Guja, C., Ionescu-Tirgoviste, C., Genetics of Type 1 Diabetes in F, Simmonds, M. J., Heward, J. M., Gough, S. C., Wellcome Trust Case Control C, Dunger, D. B., Wicker, L. S., & Clayton, D. G. (2007). Robust associations of four new chromosome regions from genome-wide analyses of type 1 diabetes. *Nature Genetics, 39*, 857–864.
10. Zeggini, E., Scott, L. J., Saxena, R., Voight, B. F., Marchini, J. L., Hu, T., de Bakker, P. I., Abecasis, G. R., Almgren, P., Andersen, G., Ardlie, K., Bostrom, K. B., Bergman, R. N., Bonnycastle, L. L., Borch-Johnsen, K., Burtt, N. P., Chen, H., Chines, P. S., Daly, M. J., Deodhar, P., Ding, C. J., Doney, A. S., Duren, W. L., Elliott, K. S., Erdos, M. R., Frayling,

T. M., Freathy, R. M., Gianniny, L., Grallert, H., Grarup, N., Groves, C. J., Guiducci, C., Hansen, T., Herder, C., Hitman, G. A., Hughes, T. E., Isomaa, B., Jackson, A. U., Jorgensen, T., Kong, A., Kubalanza, K., Kuruvilla, F. G., Kuusisto, J., Langenberg, C., Lango, H., Lauritzen, T., Li, Y., Lindgren, C. M., Lyssenko, V., Marvelle, A. F., Meisinger, C., Midthjell, K., Mohlke, K. L., Morken, M. A., Morris, A. D., Narisu, N., Nilsson, P., Owen, K. R., Palmer, C. N., Payne, F., Perry, J. R., Pettersen, E., Platou, C., Prokopenko, I., Qi, L., Qin, L., Rayner, N. W., Rees, M., Roix, J. J., Sandbaek, A., Shields, B., Sjogren, M., Steinthorsdottir, V., Stringham, H. M., Swift, A. J., Thorleifsson, G., Thorsteinsdottir, U., Timpson, N. J., Tuomi, T., Tuomilehto, J., Walker, M., Watanabe, R. M., Weedon, M. N., Willer, C. J., Wellcome Trust Case Control C, Illig, T., Hveem, K., Hu, F. B., Laakso, M., Stefansson, K., Pedersen, O., Wareham, N. J., Barroso, I., Hattersley, A. T., Collins, F. S., Groop, L., McCarthy, M. I., Boehnke, M., & Altshuler, D. (2008). Meta-analysis of genome-wide association data and large-scale replication identifies additional susceptibility loci for type 2 diabetes. *Nature Genetics, 40*, 638–645.
11. Gieger, C., Geistlinger, L., Altmaier, E., Hrabe de Angelis, M., Kronenberg, F., Meitinger, T., Mewes, H. W., Wichmann, H. E., Weinberger, K. M., Adamski, J., Illig, T., & Suhre, K. (2008). Genetics meets metabolomics: A genome-wide association study of metabolite profiles in human serum. *PLoS Genetics, 4*, e1000282.
12. Gates, S. C., & Sweeley, C. C. (1978). Quantitative metabolic profiling based on gas chromatography. *Clinical Chemistry, 24*, 1663–1673.
13. Gohlke, R. S., & McLafferty, F. W. (1993). Early gas chromatography/mass spectrometry. *Journal of the American Society for Mass Spectrometry, 4*, 367–371.
14. Kell, D. B., & Oliver, S. G. (2016). The metabolome 18 years on: A concept comes of age. *Metabolomics, 12*, 148.
15. Rosato, A., Tenori, L., Cascante, M., De Atauri Carulla, P. R., Martins Dos Santos, V. A. P., & Saccenti, E. (2018). From correlation to causation: Analysis of metabolomics data using systems biology approaches. *Metabolomics, 14*, 37.
16. Zamboni, N., Saghatelian, A., & Patti, G. J. (2015). Defining the metabolome: Size, flux, and regulation. *Molecular Cell, 58*, 699–706.
17. Horning, E. C., & Horning, M. G. (1971). Metabolic profiles: Gas-phase methods for analysis of metabolites. *Clinical Chemistry, 17*, 802–809.
18. Mamer, O. A., Crawhall, J. C., & Tjoa, S. S. (1971). The identification of urinary acids by coupled gas chromatography-mass spectrometry. *Clinica Chimica Acta, 32*, 171–184.
19. Pauling, L., Robinson, A. B., Teranishi, R., & Cary, P. (1971). Quantitative analysis of urine vapor and breath by gas-liquid partition chromatography. *Proceedings of the National Academy of Sciences of the United States of America, 68*, 2374–2376.
20. Hoult, D. I., Busby, S. J., Gadian, D. G., Radda, G. K., Richards, R. E., & Seeley, P. J. (1974). Observation of tissue metabolites using 31P nuclear magnetic resonance. *Nature, 252*, 285–287.
21. Radda, G. K., & Seeley, P. J. (1979). Recent studies on cellular metabolism by nuclear magnetic resonance. *Annual Review of Physiology, 41*, 749–769.
22. Dunn, W. B., Broadhurst, D., Begley, P., Zelena, E., Francis-McIntyre, S., Anderson, N., Brown, M., Knowles, J. D., Halsall, A., Haselden, J. N., Nicholls, A. W., Wilson, I. D., Kell, D. B., Goodacre, R., & Human Serum Metabolome Consortium. (2011). Procedures for large-scale metabolic profiling of serum and plasma using gas chromatography and liquid chromatography coupled to mass spectrometry. *Nature Protocols, 6*, 1060–1083.
23. Wollenberger, A., Ristau, O., & Schoffa, G. (1960). A simple technic for extremely rapid freezing of large pieces of tissue. *Pflügers Archiv für die Gesamte Physiologie des Menschen und der Tiere, 270*, 399–412.
24. Haid, M., Muschet, C., Wahl, S., Romisch-Margl, W., Prehn, C., Moller, G., & Adamski, J. (2018). Long-term stability of human plasma metabolites during storage at −80 degrees C. *Journal of Proteome Research, 17*, 203–211.

25. Issaq, H. J., Van, Q. N., Waybright, T. J., Muschik, G. M., & Veenstra, T. D. (2009). Analytical and statistical approaches to metabolomics research. *Journal of Separation Science, 32*, 2183–2199.
26. Ellis, D. I., Dunn, W. B., Griffin, J. L., Allwood, J. W., & Goodacre, R. (2007). Metabolic fingerprinting as a diagnostic tool. *Pharmacogenomics, 8*, 1243–1266.
27. Illig, T., Gieger, C., Zhai, G., Romisch-Margl, W., Wang-Sattler, R., Prehn, C., Altmaier, E., Kastenmuller, G., Kato, B. S., Mewes, H. W., Meitinger, T., de Angelis, M. H., Kronenberg, F., Soranzo, N., Wichmann, H. E., Spector, T. D., Adamski, J., & Suhre, K. (2010). A genome-wide perspective of genetic variation in human metabolism. *Nature Genetics, 42*, 137–141.
28. Griffiths, W. J., Karu, K., Hornshaw, M., Woffendin, G., & Wang, Y. (2007). Metabolomics and metabolite profiling: Past heroes and future developments. *European Journal of Mass Spectrometry (Chichester), 13*, 45–50.
29. Zhao, X., Fritsche, J., Wang, J., Chen, J., Rittig, K., Schmitt-Kopplin, P., Fritsche, A., Haring, H. U., Schleicher, E. D., Xu, G., & Lehmann, R. (2010). Metabonomic fingerprints of fasting plasma and spot urine reveal human pre-diabetic metabolic traits. *Metabolomics, 6*, 362–374.
30. Lawton, K. A., Berger, A., Mitchell, M., Milgram, K. E., Evans, A. M., Guo, L., Hanson, R. W., Kalhan, S. C., Ryals, J. A., & Milburn, M. V. (2008). Analysis of the adult human plasma metabolome. *Pharmacogenomics, 9*, 383–397.
31. Ohta, T., Masutomi, N., Tsutsui, N., Sakairi, T., Mitchell, M., Milburn, M. V., Ryals, J. A., Beebe, K. D., & Guo, L. (2009). Untargeted metabolomic profiling as an evaluative tool of fenofibrate-induced toxicology in Fischer 344 male rats. *Toxicologic Pathology, 37*, 521–535.
32. Sreekumar, A., Poisson, L. M., Rajendiran, T. M., Khan, A. P., Cao, Q., Yu, J., Laxman, B., Mehra, R., Lonigro, R. J., Li, Y., Nyati, M. K., Ahsan, A., Kalyana-Sundaram, S., Han, B., Cao, X., Byun, J., Omenn, G. S., Ghosh, D., Pennathur, S., Alexander, D. C., Berger, A., Shuster, J. R., Wei, J. T., Varambally, S., Beecher, C., & Chinnaiyan, A. M. (2009). Metabolomic profiles delineate potential role for sarcosine in prostate cancer progression. *Nature, 457*, 910–914.
33. Azur, M. J., Stuart, E. A., Frangakis, C., & Leaf, P. J. (2011). Multiple imputation by chained equations: What is it and how does it work? *International Journal of Methods in Psychiatric Research, 20*, 40–49.
34. Croghan, C. W., & Egeghy, P. P. (2003). *Methods of dealing with values below the limit of detection using SAS*. St. Petersburg, FL: Southeastern SAS User Group.
35. Gromski, P. S., Xu, Y., Kotze, H. L., Correa, E., Ellis, D. I., Armitage, E. G., Turner, M. L., & Goodacre, R. (2014). Influence of missing values substitutes on multivariate analysis of metabolomics data. *Metabolites, 4*, 433–452.
36. Hrydziuszko, O., & Viant, M. R. (2012). Missing values in mass spectrometry based metabolomics: An undervalued step in the data processing pipeline. *Metabolomics, 8*, 161–174.
37. Lubin, J. H., Colt, J. S., Camann, D., Davis, S., Cerhan, J. R., Severson, R. K., Bernstein, L., & Hartge, P. (2004). Epidemiologic evaluation of measurement data in the presence of detection limits. *Environmental Health Perspectives, 112*, 1691–1696.
38. Xia, J., Psychogios, N., Young, N., & Wishart, D. S. (2009). MetaboAnalyst: A web server for metabolomic data analysis and interpretation. *Nucleic Acids Research, 37*, W652–W660.
39. Brereton, R. G., & Lloyd, G. R. (2014). Partial least squares discriminant analysis: Taking the magic away. *Journal of Chemometrics, 28*, 213–225.
40. Efron, B., & Tibshirani, R. (1986). Bootstrap methods for standard errors, confidence intervals, and other measures of statistical accuracy. *Statistical Science, 1*, 54–75.
41. Eriksson, L., Trygg, J., & Wold, S. (2014). A chemometrics toolbox based on projections and latent variables. *Journal of Chemometrics, 28*, 332–346.
42. Grissa, D., Petera, M., Brandolini, M., Napoli, A., Comte, B., & Pujos-Guillot, E. (2016). Feature selection methods for early predictive biomarker discovery using untargeted metabolomic data. *Frontiers in Molecular Biosciences, 3*, 30.

43. Gromski, P. S., Muhamadali, H., Ellis, D. I., Xu, Y., Correa, E., Turner, M. L., & Goodacre, R. (2015). A tutorial review: Metabolomics and partial least squares-discriminant analysis – A marriage of convenience or a shotgun wedding. *Analytica Chimica Acta, 879*, 10–23.
44. Gromski, P. S., Xu, Y., Correa, E., Ellis, D. I., Turner, M. L., & Goodacre, R. (2014). A comparative investigation of modern feature selection and classification approaches for the analysis of mass spectrometry data. *Analytica Chimica Acta, 829*, 1–8.
45. Hastie, T., Tibshirani, R., & Friedman, J. (2009). *The elements of statistical learning*. New York: Springer.
46. Jolliffe, I. T. (2002). *Principal component analysis*. New York: Springer.
47. Kastenmuller, G., Romisch-Margl, W., Wagele, B., Altmaier, E., & Suhre, K. (2011). metaP-server: A web-based metabolomics data analysis tool. *Journal of Biomedicine and Biotechnology, 2011*, 839862.
48. Ren, S., Hinzman, A. A., Kang, E. L., Szczesniak, R. D., & Lu, L. J. (2015). Computational and statistical analysis of metabolomics data. *Metabolomics, 11*, 1492–1513.
49. Saccenti, E., Hoefsloot, H. C. J., Smilde, A. K., Westerhuis, J. A., & Hendriks, M. M. W. B. (2014). Reflections on univariate and multivariate analysis of metabolomics data. *Metabolomics, 10*, 361–374.
50. Shapiro, S. S., & Wilk, M. B. (1965). An analysis of variance test for normality (complete samples). *Biometrika, 52*, 591–611.
51. Tautenhahn, R., Patti, G. J., Rinehart, D., & Siuzdak, G. (2012). XCMS online: A web-based platform to process untargeted metabolomic data. *Analytical Chemistry, 84*, 5035–5039.
52. Vinaixa, M., Samino, S., Saez, I., Duran, J., Guinovart, J. J., & Yanes, O. (2012). A guideline to univariate statistical analysis for LC/MS-based untargeted metabolomics-derived data. *Metabolites, 2*, 775–795.
53. Wold, S., Sjöström, M., & Eriksson, L. (2001). PLS-regression: A basic tool of chemometrics. *Chemometrics and Intelligent Laboratory Systems, 58*, 109–130.
54. Xia, J., Sinelnikov, I. V., Han, B., & Wishart, D. S. (2015). MetaboAnalyst 3.0 – making metabolomics more meaningful. *Nucleic Acids Research, 43*, W251–W257.
55. Brereton, R. G. (2006). Consequences of sample size, variable selection, and model validation and optimisation, for predicting classification ability from analytical data. *TrAC Trends in Analytical Chemistry, 25*, 1103–1111.
56. Esbensen, K. H., & Geladi, P. (2010). Principles of proper validation: Use and abuse of re-sampling for validation. *Journal of Chemometrics, 24*, 168–187.
57. Trygg, J., & Wold, S. (2002). Orthogonal projections to latent structures (O-PLS). *Journal of Chemometrics, 16*, 119–128.
58. Westerhuis, J. A., Hoefsloot, H. C. J., Smit, S., Vis, D. J., Smilde, A. K., van Velzen, J. J., van Duijnhoven, J. P. M., & van Dorsten, F. A. (2008). Assessment of PLSDA cross validation. *Metabolomics, 4*, 81–89.
59. Xia, J., Broadhurst, D. I., Wilson, M., & Wishart, D. S. (2013). Translational biomarker discovery in clinical metabolomics: An introductory tutorial. *Metabolomics, 9*, 280–299.
60. Vouk, K., Hevir, N., Ribic-Pucelj, M., Haarpaintner, G., Scherb, H., Osredkar, J., Moller, G., Prehn, C., Rizner, T. L., & Adamski, J. (2012). Discovery of phosphatidylcholines and sphingomyelins as biomarkers for ovarian endometriosis. *Human Reproduction, 27*, 2955–2965.
61. Chong, J., Soufan, O., Li, C., Caraus, I., Li, S., Bourque, G., Wishart, D. S., & Xia, J. (2018). MetaboAnalyst 4.0: Towards more transparent and integrative metabolomics analysis. *Nucleic Acids Research, 46*, W486–W494.
62. Zurlo, F., Larson, K., Bogardus, C., & Ravussin, E. (1990). Skeletal muscle metabolism is a major determinant of resting energy expenditure. *The Journal of Clinical Investigation, 86*, 1423–1427.
63. Hargreaves, M. (2000). Skeletal muscle metabolism during exercise in humans. *Clinical and Experimental Pharmacology and Physiology, 27*, 225–228.

64. Shearer, J., & Weljie, A. (2014). Biomarkers of skeletal muscle regulation, metabolism and dysfunction. In: Jones, O. A. H. (Ed.), *Metabolomics and systems biology in human health and medicine* (pp. 157–170). CABI.
65. Hadrevi, J., Ghafouri, B., Sjors, A., Antti, H., Larsson, B., Crenshaw, A. G., Gerdle, B., & Hellstrom, F. (2013). Comparative metabolomics of muscle interstitium fluid in human trapezius myalgia: an in vivo microdialysis study. *European Journal of Applied Physiology, 113*, 2977–2989.
66. Zhang, J., Bhattacharyya, S., Hickner, R. C., Light, A. R., Lambert, C. J., Gale, B. K., Fiehn, O., & Adams, S. H. (2019). Skeletal muscle interstitial fluid metabolomics at rest and associated with an exercise bout: Application in rats and humans. *American Journal of Physiology Endocrinology and Metabolism, 316*, E43–E53.
67. Ivanisevic, J., Elias, D., Deguchi, H., Averell, P. M., Kurczy, M., Johnson, C. H., Tautenhahn, R., Zhu, Z., Watrous, J., Jain, M., Griffin, J., Patti, G. J., & Siuzdak, G. (2015). Arteriovenous blood metabolomics: A readout of intra-tissue metabostasis. *Scientific Reports, 5*, 12757.
68. Dotzert, M. S., Murray, M. R., McDonald, M. W., Olver, T. D., Velenosi, T. J., Hennop, A., Noble, E. G., Urquhart, B. L., & Melling, C. W. (2016). Metabolomic response of skeletal muscle to aerobic exercise training in insulin resistant type 1 diabetic rats. *Scientific Reports, 6*, 26379.
69. Goto-Inoue, N., Yamada, K., Inagaki, A., Furuichi, Y., Ogino, S., Manabe, Y., Setou, M., & Fujii, N. L. (2013). Lipidomics analysis revealed the phospholipid compositional changes in muscle by chronic exercise and high-fat diet. *Scientific Reports, 3*, 3267.
70. Ilaiwy, A., Quintana, M. T., Bain, J. R., Muehlbauer, M. J., Brown, D. I., Stansfield, W. E., & Willis, M. S. (2016). Cessation of biomechanical stretch model of C2C12 cells models myocyte atrophy and anaplerotic changes in metabolism using non-targeted metabolomics analysis. *The International Journal of Biochemistry and Cell Biology, 79*, 80–92.
71. Saoi, M., Percival, M., Nemr, C., Li, A., Gibala, M., & Britz-McKibbin, P. (2019). Characterization of the human skeletal muscle metabolome for elucidating the mechanisms of bicarbonate ingestion on strenuous interval exercise. *Analytical Chemistry, 91*, 4709–4718.
72. Sato, S., Parr, E. B., Devlin, B. L., Hawley, J. A., & Sassone-Corsi, P. (2018). Human metabolomics reveal daily variations under nutritional challenges specific to serum and skeletal muscle. *Molecular Metabolism, 16*, 1–11.
73. Starnes, J. W., Parry, T. L., O'Neal, S. K., Bain, J. R., Muehlbauer, M. J., Honcoop, A., Ilaiwy, A., Christopher, P. M., Patterson, C., & Willis, M. S. (2017). Exercise-induced alterations in skeletal muscle, heart, liver, and serum metabolome identified by non-targeted metabolomics analysis. *Metabolites, 7*, 40.
74. White, P. J., Lapworth, A. L., An, J., Wang, L., McGarrah, R. W., Stevens, R. D., Ilkayeva, O., George, T., Muehlbauer, M. J., Bain, J. R., Trimmer, J. K., Brosnan, M. J., Rolph, T. P., & Newgard, C. B. (2016). Branched-chain amino acid restriction in Zucker-fatty rats improves muscle insulin sensitivity by enhancing efficiency of fatty acid oxidation and acyl-glycine export. *Molecular Metabolism, 5*, 538–551.
75. Abdullah, M., Kornegay, J. N., Honcoop, A., Parry, T. L., Balog-Alvarez, C. J., O'Neal, S. K., Bain, J. R., Muehlbauer, M. J., Newgard, C. B., Patterson, C., & Willis, M. S. (2017). Non-targeted metabolomics analysis of golden retriever muscular dystrophy-affected muscles reveals alterations in arginine and proline metabolism, and elevations in glutamic and oleic acid in vivo. *Metabolites, 7*(3), 38.
76. Buzkova, J., Nikkanen, J., Ahola, S., Hakonen, A. H., Sevastianova, K., Hovinen, T., Yki-Jarvinen, H., Pietilainen, K. H., Lonnqvist, T., Velagapudi, V., Carroll, C. J., & Suomalainen, A. (2018). Metabolomes of mitochondrial diseases and inclusion body myositis patients: Treatment targets and biomarkers. *EMBO Molecular Medicine, 10*, e9091.
77. Griffin, J. L., & Des, R. C. (2009). Applications of metabolomics and proteomics to the mdx mouse model of Duchenne muscular dystrophy: Lessons from downstream of the transcriptome. *Genome Medicine, 1*, 32.

78. Du, T., Lin, Z., Xie, Y., Ye, X., Tu, C., Jin, K., Xie, J., & Shen, Y. (2018). Metabolic profiling of femoral muscle from rats at different periods of time after death. *PLoS One, 13*, e0203920.
79. Fazelzadeh, P., Hangelbroek, R. W., Tieland, M., de Groot, L. C., Verdijk, L. B., van Loon, L. J., Smilde, A. K., Alves, R. D., Vervoort, J., Muller, M., van Duynhoven, J. P., & Boekschoten, M. V. (2016). The muscle metabolome differs between healthy and frail older adults. *Journal of Proteome Research, 15*, 499–509.
80. Garvey, S. M., Dugle, J. E., Kennedy, A. D., McDunn, J. E., Kline, W., Guo, L., Guttridge, D. C., Pereira, S. L., & Edens, N. K. (2014). Metabolomic profiling reveals severe skeletal muscle group-specific perturbations of metabolism in aged FBN rats. *Biogerontology, 15*, 217–232.
81. Houtkooper, R. H., Argmann, C., Houten, S. M., Canto, C., Jeninga, E. H., Andreux, P. A., Thomas, C., Doenlen, R., Schoonjans, K., & Auwerx, J. (2011). The metabolic footprint of aging in mice. *Scientific Reports, 1*, 134.
82. Jang, C., Oh, S. F., Wada, S., Rowe, G. C., Liu, L., Chan, M. C., Rhee, J., Hoshino, A., Kim, B., Ibrahim, A., Baca, L. G., Kim, E., Ghosh, C. C., Parikh, S. M., Jiang, A., Chu, Q., Forman, D. E., Lecker, S. H., Krishnaiah, S., Rabinowitz, J. D., Weljie, A. M., Baur, J. A., Kasper, D. L., & Arany, Z. (2016). A branched-chain amino acid metabolite drives vascular fatty acid transport and causes insulin resistance. *Nature Medicine, 22*, 421–426.
83. Montgomery, M. K., Brown, S. H. J., Mitchell, T. W., Coster, A. C. F., Cooney, G. J., & Turner, N. (2017). Association of muscle lipidomic profile with high-fat diet-induced insulin resistance across five mouse strains. *Scientific Reports, 7*, 13914.
84. Roberts, L. D., Bostrom, P., O'Sullivan, J. F., Schinzel, R. T., Lewis, G. D., Dejam, A., Lee, Y. K., Palma, M. J., Calhoun, S., Georgiadi, A., Chen, M. H., Ramachandran, V. S., Larson, M. G., Bouchard, C., Rankinen, T., Souza, A. L., Clish, C. B., Wang, T. J., Estall, J. L., Soukas, A. A., Cowan, C. A., Spiegelman, B. M., & Gerszten, R. E. (2014). beta-Aminoisobutyric acid induces browning of white fat and hepatic beta-oxidation and is inversely correlated with cardiometabolic risk factors. *Cell Metabolism, 19*, 96–108.
85. Tonks, K. T., Coster, A. C., Christopher, M. J., Chaudhuri, R., Xu, A., Gagnon-Bartsch, J., Chisholm, D. J., James, D. E., Meikle, P. J., Greenfield, J. R., & Samocha-Bonet, D. (2016). Skeletal muscle and plasma lipidomic signatures of insulin resistance and overweight/obesity in humans. *Obesity (Silver Spring), 24*, 908–916.
86. Wood, P., & Shirley, N. (2013). Lipidomics analysis of postmortem interval: Preliminary evaluation of human skeletal muscle. *Metabolomics, 3*, 127–129.
87. Aguer, C., Piccolo, B. D., Fiehn, O., Adams, S. H., & Harper, M. E. (2017). A novel amino acid and metabolomics signature in mice overexpressing muscle uncoupling protein 3. *The FASEB Journal, 31*, 814–827.
88. An, J., Muoio, D. M., Shiota, M., Fujimoto, Y., Cline, G. W., Shulman, G. I., Koves, T. R., Stevens, R., Millington, D., & Newgard, C. B. (2004). Hepatic expression of malonyl-CoA decarboxylase reverses muscle, liver and whole-animal insulin resistance. *Nature Medicine, 10*, 268–274.
89. Cheng, K. K., Akasaki, Y., Lecommandeur, E., Lindsay, R. T., Murfitt, S., Walsh, K., & Griffin, J. L. (2015). Metabolomic analysis of akt1-mediated muscle hypertrophy in models of diet-induced obesity and age-related fat accumulation. *Journal of Proteome Research, 14*, 342–352.
90. Choi, C. S., Befroy, D. E., Codella, R., Kim, S., Reznick, R. M., Hwang, Y. J., Liu, Z. X., Lee, H. Y., Distefano, A., Samuel, V. T., Zhang, D., Cline, G. W., Handschin, C., Lin, J., Petersen, K. F., Spiegelman, B. M., & Shulman, G. I. (2008). Paradoxical effects of increased expression of PGC-1alpha on muscle mitochondrial function and insulin-stimulated muscle glucose metabolism. *Proceedings of the National Academy of Sciences of the United States of America, 105*, 19926–19931.
91. Koves, T. R., Li, P., An, J., Akimoto, T., Slentz, D., Ilkayeva, O., Dohm, G. L., Yan, Z., Newgard, C. B., & Muoio, D. M. (2005). Peroxisome proliferator-activated receptor-gamma co-activator 1alpha-mediated metabolic remodeling of skeletal myocytes mimics exercise

training and reverses lipid-induced mitochondrial inefficiency. *The Journal of Biological Chemistry, 280*, 33588–33598.

92. Park, S. M., Byeon, S. K., Lee, H., Sung, H., Kim, I. Y., Seong, J. K., & Moon, M. H. (2017). Lipidomic analysis of skeletal muscle tissues of p53 knockout mice by nUPLC-ESI-MS/MS. *Scientific Reports, 7*, 3302.
93. Roberts, L. D., Hassall, D. G., Winegar, D. A., Haselden, J. N., Nicholls, A. W., & Griffin, J. L. (2009). Increased hepatic oxidative metabolism distinguishes the action of Peroxisome proliferator-activated receptor delta from Peroxisome proliferator-activated receptor gamma in the ob/ob mouse. *Genome Medicine, 1*, 115.
94. Wong, K. E., Mikus, C. R., Slentz, D. H., Seiler, S. E., DeBalsi, K. L., Ilkayeva, O. R., Crain, K. I., Kinter, M. T., Kien, C. L., Stevens, R. D., & Muoio, D. M. (2015). Muscle-specific overexpression of PGC-1alpha does not augment metabolic improvements in response to exercise and caloric restriction. *Diabetes, 64*, 1532–1543.
95. Wu, C. L., Satomi, Y., & Walsh, K. (2017). RNA-seq and metabolomic analyses of Akt1-mediated muscle growth reveals regulation of regenerative pathways and changes in the muscle secretome. *BMC Genomics, 18*, 181.
96. York, B., Reineke, E. L., Sagen, J. V., Nikolai, B. C., Zhou, S., Louet, J. F., Chopra, A. R., Chen, X., Reed, G., Noebels, J., Adesina, A. M., Yu, H., Wong, L. J., Tsimelzon, A., Hilsenbeck, S., Stevens, R. D., Wenner, B. R., Ilkayeva, O., Xu, J., Newgard, C. B., & O'Malley, B. W. (2012). Ablation of steroid receptor coactivator-3 resembles the human CACT metabolic myopathy. *Cell Metabolism, 15*, 752–763.
97. Koves, T. R., Ussher, J. R., Noland, R. C., Slentz, D., Mosedale, M., Ilkayeva, O., Bain, J., Stevens, R., Dyck, J. R., Newgard, C. B., Lopaschuk, G. D., & Muoio, D. M. (2008). Mitochondrial overload and incomplete fatty acid oxidation contribute to skeletal muscle insulin resistance. *Cell Metabolism, 7*, 45–56.
98. Dyar, K. A., Lutter, D., Artati, A., Ceglia, N. J., Liu, Y., Armenta, D., Jastroch, M., Schneider, S., de Mateo, S., Cervantes, M., Abbondante, S., Tognini, P., Orozco-Solis, R., Kinouchi, K., Wang, C., Swerdloff, R., Nadeef, S., Masri, S., Magistretti, P., Orlando, V., Borrelli, E., Uhlenhaut, N. H., Baldi, P., Adamski, J., Tschop, M. H., Eckel-Mahan, K., & Sassone-Corsi, P. (2018). Atlas of circadian metabolism reveals system-wide coordination and communication between clocks. *Cell, 174*, 1571–1585.e11.
99. Abbondante, S., Eckel-Mahan, K. L., Ceglia, N. J., Baldi, P., & Sassone-Corsi, P. (2016). Comparative circadian metabolomics reveal differential effects of nutritional challenge in the serum and liver. *The Journal of Biological Chemistry, 291*, 2812–2828.
100. Adamovich, Y., Rousso-Noori, L., Zwighaft, Z., Neufeld-Cohen, A., Golik, M., Kraut-Cohen, J., Wang, M., Han, X., & Asher, G. (2014). Circadian clocks and feeding time regulate the oscillations and levels of hepatic triglycerides. *Cell Metabolism, 19*, 319–330.
101. Aviram, R., Manella, G., Kopelman, N., Neufeld-Cohen, A., Zwighaft, Z., Elimelech, M., Adamovich, Y., Golik, M., Wang, C., Han, X., & Asher, G. (2016). Lipidomics analyses reveal temporal and spatial lipid organization and uncover daily oscillations in intracellular organelles. *Molecular Cell, 62*, 636–648.
102. Dyar, K. A., & Eckel-Mahan, K. L. (2017). Circadian metabolomics in time and space. *Frontiers in Neuroscience, 11*, 369.
103. Eckel-Mahan, K. L., Patel, V. R., de Mateo, S., Orozco-Solis, R., Ceglia, N. J., Sahar, S., Dilag-Penilla, S. A., Dyar, K. A., Baldi, P., & Sassone-Corsi, P. (2013). Reprogramming of the circadian clock by nutritional challenge. *Cell, 155*, 1464–1478.
104. Eckel-Mahan, K. L., Patel, V. R., Mohney, R. P., Vignola, K. S., Baldi, P., & Sassone-Corsi, P. (2012). Coordination of the transcriptome and metabolome by the circadian clock. *Proceedings of the National Academy of Sciences of the United States of America, 109*, 5541–5546.
105. Gooley, J. J., & Chua, E. C. (2014). Diurnal regulation of lipid metabolism and applications of circadian lipidomics. *Journal of Genetics and Genomics, 41*, 231–250.

106. Krishnaiah, S. Y., Wu, G., Altman, B. J., Growe, J., Rhoades, S. D., Coldren, F., Venkataraman, A., Olarerin-George, A. O., Francey, L. J., Mukherjee, S., Girish, S., Selby, C. P., Cal, S., Er, U., Sianati, B., Sengupta, A., Anafi, R. C., Kavakli, I. H., Sancar, A., Baur, J. A., Dang, C. V., Hogenesch, J. B., & Weljie, A. M. (2017). Clock regulation of metabolites reveals coupling between transcription and metabolism. *Cell Metabolism, 25*, 961–974.e4.
107. Sato, S., Basse, A. L., Schonke, M., Chen, S., Samad, M., Altintas, A., Laker, R. C., Dalbram, E., Barres, R., Baldi, P., Treebak, J. T., Zierath, J. R., & Sassone-Corsi, P. (2019). Time of exercise specifies the impact on muscle metabolic pathways and systemic energy homeostasis. *Cell Metabolism, 30*(1), 92–110.e4.
108. Loizides-Mangold, U., Perrin, L., Vandereycken, B., Betts, J. A., Walhin, J. P., Templeman, I., Chanon, S., Weger, B. D., Durand, C., Robert, M., Paz Montoya, J., Moniatte, M., Karagounis, L. G., Johnston, J. D., Gachon, F., Lefai, E., Riezman, H., & Dibner, C. (2017). Lipidomics reveals diurnal lipid oscillations in human skeletal muscle persisting in cellular myotubes cultured in vitro. *Proceedings of the National Academy of Sciences of the United States of America, 114*, E8565–E8574.
109. Dyar, K. A., Ciciliot, S., Wright, L. E., Bienso, R. S., Tagliazucchi, G. M., Patel, V. R., Forcato, M., Paz, M. I., Gudiksen, A., Solagna, F., Albiero, M., Moretti, I., Eckel-Mahan, K. L., Baldi, P., Sassone-Corsi, P., Rizzuto, R., Bicciato, S., Pilegaard, H., Blaauw, B., & Schiaffino, S. (2014). Muscle insulin sensitivity and glucose metabolism are controlled by the intrinsic muscle clock. *Molecular Metabolism, 3*, 29–41.
110. Dyar, K. A., Hubert, M. J., Mir, A. A., Ciciliot, S., Lutter, D., Greulich, F., Quagliarini, F., Kleinert, M., Fischer, K., Eichmann, T. O., Wright, L. E., Pena Paz, M. I., Casarin, A., Pertegato, V., Romanello, V., Albiero, M., Mazzucco, S., Rizzuto, R., Salviati, L., Biolo, G., Blaauw, B., Schiaffino, S., & Uhlenhaut, N. H. (2018). Transcriptional programming of lipid and amino acid metabolism by the skeletal muscle circadian clock. *PLoS Biology, 16*, e2005886.
111. DeFronzo, R. A., & Tripathy, D. (2009). Skeletal muscle insulin resistance is the primary defect in type 2 diabetes. *Diabetes Care, 32*(Suppl 2), S157–S163.
112. Kiens, B. (2006). Skeletal muscle lipid metabolism in exercise and insulin resistance. *Physiological Reviews, 86*, 205–243.
113. Wolfe, R. R. (2006). The underappreciated role of muscle in health and disease. *The American Journal of Clinical Nutrition, 84*, 475–482.
114. Wagenmakers, A. J. (1998). Muscle amino acid metabolism at rest and during exercise: role in human physiology and metabolism. *Exercise and Sport Sciences Reviews, 26*, 287–314.
115. Felig, P. (1975). Amino acid metabolism in man. *Annual Review of Biochemistry, 44*, 933–955.
116. Argiles, J. M., Campos, N., Lopez-Pedrosa, J. M., Rueda, R., & Rodriguez-Manas, L. (2016). Skeletal muscle regulates metabolism via interorgan crosstalk: Roles in health and disease. *Journal of the American Medical Directors Association, 17*, 789–796.
117. Pedersen, B. K., & Febbraio, M. A. (2012). Muscles, exercise and obesity: Skeletal muscle as a secretory organ. *Nature Reviews Endocrinology, 8*, 457–465.
118. Isherwood, C. M., Van der Veen, D. R., Johnston, J. D., & Skene, D. J. (2017). Twenty-four-hour rhythmicity of circulating metabolites: Effect of body mass and type 2 diabetes. *The FASEB Journal, 31*, 5557–5567.
119. Jang, C., Chen, L., & Rabinowitz, J. D. (2018). Metabolomics and isotope tracing. *Cell, 173*, 822–837.
120. Trivedi, D. K., Hollywood, K. A., & Goodacre, R. (2017). Metabolomics for the masses: The future of metabolomics in a personalized world. *New Horizons in Translational Medicine, 3*, 294–305.
121. Kennedy, A. D., Wittmann, B. M., Evans, A. M., Miller, L. A. D., Toal, D. R., Lonergan, S., Elsea, S. H., & Pappan, K. L. (2018). Metabolomics in the clinic: A review of the shared and unique features of untargeted metabolomics for clinical research and clinical testing. *Journal of Mass Spectrometry, 53*, 1143–1154.

122. Beger, R. D., Dunn, W., Schmidt, M. A., Gross, S. S., Kirwan, J. A., Cascante, M., Brennan, L., Wishart, D. S., Oresic, M., Hankemeier, T., Broadhurst, D. I., Lane, A. N., Suhre, K., Kastenmuller, G., Sumner, S. J., Thiele, I., Fiehn, O., Kaddurah-Daouk, R., & for "Precision Medicine, and Pharmacometabolomics Task Group"-Metabolomics Society Initiative. (2016). Metabolomics enables precision medicine: "A White Paper, Community Perspective". *Metabolomics, 12*, 149.
123. Hood, L. (2013). Systems biology and p4 medicine: Past, present, and future. *Rambam Maimonides Medical Journal, 4*, e0012.
124. UN. (2017). *World population prospects: The 2017 revision, key findings and advance tables, edited by United Nations DoEaSA, Population division.* New York: United Nations.
125. Friedrich, M. J. (2017). Global obesity epidemic worsening. *JAMA, 318*, 603.
126. Schiaffino, S., & Reggiani, C. (2011). Fiber types in mammalian skeletal muscles. *Physiological Reviews, 91*, 1447–1531.
127. Murgia, M., Nagaraj, N., Deshmukh, A. S., Zeiler, M., Cancellara, P., Moretti, I., Reggiani, C., Schiaffino, S., & Mann, M. (2015). Single muscle fiber proteomics reveals unexpected mitochondrial specialization. *EMBO Reports, 16*, 387–395.
128. Schiaffino, S., Reggiani, C., Kostrominova, T. Y., Mann, M., & Murgia, M. (2015). Mitochondrial specialization revealed by single muscle fiber proteomics: Focus on the Krebs cycle. *Scandinavian Journal of Medicine and Science in Sports, 25*(Suppl 4), 41–48.
129. Duncan, K. D., Fyrestam, J., & Lanekoff, I. (2019). Advances in mass spectrometry based single-cell metabolomics. *Analyst, 144*, 782–793.
130. Kim, J., & Hoppel, C. L. (2013). Comprehensive approach to the quantitative analysis of mitochondrial phospholipids by HPLC-MS. *Journal of Chromatography. B, Analytical Technologies in the Biomedical and Life Sciences, 912*, 105–114.
131. Furuichi, Y., Goto-Inoue, N., Manabe, Y., Setou, M., Masuda, K., & Fujii, N. L. (2014). Imaging mass spectrometry reveals fiber-specific distribution of acetylcarnitine and contraction-induced carnitine dynamics in rat skeletal muscles. *Biochimica et Biophysica Acta, 1837*, 1699–1706.
132. Goto-Inoue, N., Manabe, Y., Miyatake, S., Ogino, S., Morishita, A., Hayasaka, T., Masaki, N., Setou, M., & Fujii, N. L. (2012). Visualization of dynamic change in contraction-induced lipid composition in mouse skeletal muscle by matrix-assisted laser desorption/ionization imaging mass spectrometry. *Analytical and Bioanalytical Chemistry, 403*, 1863–1871.
133. Goto-Inoue, N., Morisasa, M., Machida, K., Furuichi, Y., Fujii, N. L., Miura, S., & Mori, T. (2019). Characterization of myofiber-type-specific molecules using mass spectrometry imaging. *Rapid Communications in Mass Spectrometry, 33*, 185–192.
134. Tsai, Y. H., Garrett, T. J., Carter, C. S., & Yost, R. A. (2015). Metabolomic analysis of oxidative and glycolytic skeletal muscles by matrix-assisted laser desorption/ionizationmass spectrometric imaging (MALDI MSI). *Journal of the American Society for Mass Spectrometry, 26*, 915–923.

Printed in the United States
By Bookmasters